Medal of Honor

Volume 1

Aviators of World War One

Alan E. Durkota

Color Illustrations by

Juanita Franzi   **Jean Galli**

Dedication
To Donna,
for her continuous encouragement and support throughout my many research projects.

Acknowledgments
I wish to extend my sincere thanks to the many people and institutions who were helpful throughout this project. Appearing in alphabetical order: Robert Cavanagh; The Congressional Medal of Honor Society, Mt. Pleasant, South Carolina; James Davilla; Peter Doyle; Donna Durkota for endless efforts in organizing my notes; Juanita Franzi for making the outstanding aircraft illustrations used in this book; Jean Galli for the wonderful color images created for this book; Dennis Gordon for an endless supply information and photographs; Jon Guttman for his support in providing historic data; Gretchen and Jack Herris for their endless efforts in creating books of the highest quality; Edward Milligan for research at the various archives; Nick Mladenoff for the many rare and valuable photographs; National Archives, College Park, Maryland; National Museum of Naval Aviation, Pensacola, Florida; Naval Historical Center, Washington D.C.; Bob Sheldon; Smithsonian Institution's National Air & Space Museum and staff member Russell Lee and Wesley Smith; Noel Shirley for his efforts regarding Naval History; Susan Stobiecki; Alan Toelle for help in numerous areas, but most of all regarding aircraft colors; United States Marine Corps Historical Center, Washington D.C.; United States Naval Institute, Annapolis, Maryland; Greg VanWyngarden; George Williams for needed photographs.

Alan E. Durkota

Contents

© 1998 by Alan E. Durkota and Flying Machines Press
Published in the United States by Flying Machines Press, 35 Chelsea Street, Stratford, Connecticut, 06497
Book and cover design, layout, and typesetting by John W. Herris.
Color aircraft illustrations by Juanita Franzi.
Color figure and medal illustrations by Jean Galli.
Digital scanning and image editing by John W. Herris.
Text edited by John W. Herris.
Printed and Bound in Korea.

Publisher's Cataloging-in-Publication Data
Durkota, Alan E., 1960–
Medal of Honor, Volume 1, Aviators of World War One / Alan Durkota
      p.   cm.
Includes bibliographical references.
ISBN 1-891268-03-1 (alk. paper)
1. Airplanes, Military—American—History.
2. World War, 1914–1918—Aerial operations, American.
3. World War, 1914–1918—Equipment.
UG1245.F8D38   1998
623.7′461′094409041—dc21   98-xxxxx
             CIP

Flying Machines Press

For the latest information on our books and posters, visit our web site at:

www.flying-machines.com

Tel.: (203 378-9344   Fax: (203) 380-1849

Medal of Honor

The Medal of Honor is the highest military award for valor that can be bestowed upon an individual actively serving in the Armed Forces of the United States. Established by the Federal Government to recognize individual acts of bravery during the Civil War, the Medal has remained America's supreme tribute for nearly 140 years and has been awarded to over 3,400 men, and in one instance, a woman. It is generally presented to the recipient by the President of the United States of America in the name of Congress, and for this reason it is often mistakenly called the Congressional Medal of Honor. This point is difficult to understand since there is an organization called The Congressional Medal of Honor Society. However, the only intent of the word *Congressional* in the name was to indicate this society is chartered by the United States Congress.

Not all awards have been for combat service; nearly 200 people have received the Medal of Honor for personal bravery or self-sacrifice while the United States was at peace. And others were "special legislation" awards presented by Congress to individuals for their "lifetime of service," or for "contributions in exploration."

The Navy accounted for 180 of the peacetime Medals of Honor.[1] Yet the reason for awarding them was established in the original Congressional act which created the Medal on December 21, 1861, specifying gallantry in action "and other seamanlike qualities" as the basis for the award.[2] For this reason the Navy could and did award the Medal of Honor for bravery in the line of the naval profession. Such awards recognized bravery in saving life and acts of valor performed in submarine rescues, boiler explosions, turret fires, and other types of disaster unique to the naval profession. But in each case, the individual distinguished himself with conspicuous gallantry or intrepidity, at risk of life above and beyond the call of duty.

There are three different Medal of Honor designs in use today: the original five-pointed star established in 1862 and still used by the Navy, Marine Corps, and Coast Guard; a wreath version adopted by the Army in 1904; and an altered wreath version adopted by the Air Force in 1965.

The Navy's Medal was the first decoration formally authorized by the American Government to be worn as a badge of honor.[3] For that reason, there was no precedent to guide the Navy in its design. Therefore, only days after President Lincoln signed the bill which authorized the Navy's Medal of Honor, Secretary of the Navy Gideon Wells contacted James Pollock, then director of the U.S. Mint at Philadelphia, and requested his assistance in creating an appropriate design for the award. Over the next five months, Pollock submitted seven designs with the aid of a Philadelphia silversmith firm – William Wilson and Son – before Wells finally approved one of them on May 9, 1962.

Soon after, on July 12, 1862, President Lincoln signed the bill which authorized the Army's Medal of Honor. Like Wells, Secretary of War Edwin Stanton was unimpressed by the mint's early attempts at an Army medal, noting he wanted a design that differed from the Navy's.

The Army's 1904 "Gillespie" Medal of Honor design. Four Army Air Corps aviators were awarded this Medal of Honor for acts of bravery during World War One – Erwin Bleckley, Harold Geottler, Frank Luke, Jr., and Edward Rickenbacker. Rickenbacker was the only one who survived the war. His Medal was presented to him on July 15, 1930. Rickenbacker's Medal differed by the fact his was suspended by a cravat (neck ribbon) of unique design. (Photo courtesy of The Congressional Medal of Honor Society.)

The Navy's 1913 Medal of Honor design. Chief Machinst's Mate Francis Ormsbee, USN, was awarded a medal of this design in 1918. (Photo courtesy of The Congressional Medal of Honor Society.)

The Navy's 'Tiffany cross' Medal of Honor design. Three aviators were awarded this Medal of Honor for acts of bravery during World War One – Landsman for Quartermaster Charles Hammann, USN; Gunnery Sergeant Robert Robinson, USMC; and Second Lieutenant Ralph Talbot, USMC. (Photo courtesy of The Congressional Medal of Honor Society.)

In response, the firm William Wilson and Son submitted a proposed medal in October 1862. It was accepted by Staton on November 17, 1862, although the only differences between the Army and Navy medals was the method by which each was attached to its ribbon and the top suspender brooch. The Navy's medal was suspended by means of an anchor and star symbol, while the Army's medal used an American eagle symbol, standing on crossed cannon and cannon balls with extended wings.[4] Otherwise the original, 1862, Army and Navy designs were both five-pointed 'inverted' bronze stars, attached to the left breast, by a ribbon having an upper portion of solid blue, and a lower portion divided vertically into alternating red and white stripes. There were 13 stripes total – six white and seven red – like the American flag.

Over the years, the Army and Navy have altered their Medal designs several times. The Army's second version appeared in 1896, and differed in ribbon only, having five vertical stripes (red, blue, white, blue, red) that ran the full length of the ribbon.

The Army's third version was authorized by Congress

on April 23, 1904. It is unofficially, yet affectionately, called the "Gillespie" design, after General G. L. Gillespie, who guarded it from abuse by obtaining Patent protection on November 22, 1904. Gillespie's final step in protecting the design came on December 19, 1904, when he transferred the patent to the Secretary of War of the United States of America.

This medal "as officially described is made of silver, heavily electroplated in gold."(5) The medal is a five-pointed 'inverted' star, tipped with trefoils. At the center of the star appears the helmet-covered head of Minerva, the Roman goddess of wisdom and the arts. Surrounding this figure in a circular pattern are the words *UNITED STATES OF AMERICA*. Interconnecting the five arms of the star is a laurel wreath enameled in green. Likewise, oak leaves carried on the base of each arm are enameled in green. This medal's pale blue ribbon carries 13 white stars, and is attached to the left breast by a hidden suspender brooch. The lower end of the ribbon was attached by a ring to an American eagle that made up part of the decorative horizontal bar. The bar itself carried the word *VALOR* and was attached to two arms of the star by smaller rings. This Medal is still in use today. However, the breast ribbon has been changed to a neck ribbon.

The Navy's first design alteration occurred in 1913, and like the Army it was limited to the ribbon. In detail the medal was a five-pointed 'inverted' bronze stars, tipped with trefoils, each arm of the star decorated with oak leaves and laurel. At the center of the star appears Minerva, symbolizing the United Sates, repulsing discord. Surrounding these figures in a circular pattern are 34 stars, representing the number of States in 1862. This medal's pale blue ribbon carries 13 white stars, and is attached to the left breast by a hollow-centered suspender brooch. The lower end of the ribbon was attached by a ring to a anchor symbol, which in turn was attached to two arms of the star by smaller rings.

The Navy's third version was authorized by Congress on February 4, 1919, and issued until 1942. Generally called the 'Tiffany cross,' this Medal is a gold patee cross with straight-sided arms that widen as they extended outward. Each arm carries an anchor, and the arms meet at a central octagonal medallion which carries in a circular pattern the legend *UNITED STATES NAVY • 1917–1918*. The center of the octagon shows the Great Seal of the United Sates – an American eagle with its shield of arms. A laurel wreath, showing between the extending arms, encircles the central medallion. This medal was attached to the left breast by a gold suspender brooch which carried the word *VALOUR* and was suspended by a pale blue ribbon with 13 white stars.

When the Navy's 'Tiffany cross' design was approved by Congressional Act on February 4, 1919, it was intended as an award for any person in the naval service of the United Sates who, while in action involving "actual conflict' with the enemy, distinguished himself conspicuously by gallantry and intrepidity at the risk of his life above and beyond the call of duty. Yet at the same time, the 1913 version was retained as an award for non-combat service. The Congressional act of August 7, 1942, eliminated the

Patrick McGunigal, Shipfitter First Class, USN. McGunigal was awarded The Great War's first Medal of Honor for rescuing a balloonist. Although McGunical was not an aviator, his Medal was the first involving aviation. (Photo courtsey U.S. Naval Historical Center.)

Navy's Medal for non-combat service and re-instituted the 1913 design for the Navy, Marine Corps, and Coast Guard. However, the breast ribbon was changed to a neck ribbon design.

The Air Force's Medal of Honor design was adopted in 1965. This medal is a five-pointed 'inverted' bronze star tipped with trefoils, each arm of the star decorated with oak leaves and laurel. Interconnecting the five arms of the star is a laurel wreath enameled in green. Likewise, oak leaves carried on the base of each arm are enameled in green. At the center of the star appears the head of the Statue of Liberty. Surrounding this figure in a circular pattern are 34 stars representing the number of States in 1862. The medal is suspended from a pale blue neck ribbon which carries 13 white stars. The attachment point is a rectangular panel which carries the word *VALOR*. Below the panel is a trophy of thunderbolts and a pair of pilot's wings.

The First Aviation-Related Medal of Honor

For acts of bravery during World War One, 123 individuals were awarded the Medal of Honor. The first was bestowed upon Patrick McGunigal, Shipfitter First Class, USN. His Medal is made more interesting, by the fact McGunigal's award was the first based on an event that involved aviation.

McGunigal was not an aviator, but was attached to the Navy's armored cruiser USS *Huntington*, which had been fitted with a kite balloon for aerial observation. McGunical's Medal of Honor citation best describes the event that occured during the ship's first Atlantic convoy: "For extraordinary heroism while attached to the *Huntington*. On the morning of 17 September 1917, while the U.S.S. *Huntington* was passing through the war zone, a kite balloon was sent up with Lt.(j.g.) H.W. Hoyt, U.S.

Navy, as observer. When the balloon was about 400 feet in the air, the temperature suddenly dropped, causing the balloon to descend about 200 feet, when it was struck by a squall. The balloon was hauled to the ship's side, but the basket trailed in the water and the pilot was submerged. McGunigal, with great daring, climbed down the side of the ship, jumped to the ropes leading to the basket, and cleared the tangle enough to get the pilot out of them. He then helped the pilot to get clear, put a bowline around him, and enabled him to be hauled to the deck. A bowline was lowered to McGunigal and he was taken safely aboad."

This book is the first volume in a series that will detail the careers of each aviator awarded the Medal of Honor. This volume is devoted to the eight aviators who were awarded the Medal in World War One.

The armoured cruiser USS *Huntington* shown with its kite balloon being lofted at Pensacola, Florida, June 23, 1917. A Curtiss N-9H is aboard and another seaplane is flying overhead. (Photo courtsey National Archives.)

Erwin R. Bleckley and Harold E. Goettler

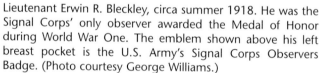

Lieutenant Erwin R. Bleckley, circa summer 1918. He was the Signal Corps' only observer awarded the Medal of Honor during World War One. The emblem shown above his left breast pocket is the U.S. Army's Signal Corps Observers Badge. (Photo courtesy George Williams.)

Lieutenant Harold E. Goettler displays his aviator's wings above his left breast pocket. He was the Signal Corps' only two-seater pilot awarded the Medal of Honor in World War One. (Photo courtesy George Williams.)

After a series of conferences between Marshal Foch, the Allied Commander-in-Chief, and General Pershing, Commander-in-Chief of the American Expeditionary Forces, it was agreed the American First Army could carry-out the St. Mihiel Offensive, but its objectives would be limited in scope so that the Americans could undertake another major offensive about two weeks later on the front between the Meuse River and the Argonne Forest. This agreement put a great burden on the First Army, for under it American troops were expected to carry to conclusion an important offensive at St. Mihiel which was scheduled to start on September 12, to concentrate an enormous force on the Meuse-Argonne front, and to then initiate a still greater operation there on September 26.

To achieve these goals in a brief span of two weeks seemed nearly impossible. Yet, from Foch's point of view, what it offered the Allies outweighed any concerns the Americans had, for the attack was to be directed against the strategic railroad center located in the vicinity of the Meuse River. The Allies believed that by gaining control of this area, the German armies would be isolated. And once this was accomplished, the enemy would not be able to maintain separate forces in France and Belgium.

To support both major offensives, the American First Army Air Service was organized in the summer of 1918, with the idea of performed aerial observation for ground troops. A wide variety of missions were carried out daily from dawn to dusk once the ground attacks began, with special flights being flown whenever the tactical situation warranted it. In fact, nothing short of heavy rain or fog was allowed to interfere with the daily aerial observation, bombardment, and pursuit missions.

For the Corps Observation Groups, perhaps the most dangerous missions were the infantry contact patrols, which required crews to air-drop desperately needed

Harold E. Goettler shown in pre-war dress, June 1913. He attended the University of Illinois prior to America's declaration of war. (Photo courtesy Dennis Gordon.)

Lieutenant Daniel Morse, Jr., C.O. of the 50th Aero Squadron. He served with the 1st Aero Squadron from September 1917 to July 1918 before joining the 50th. Morse is shown standing next to the squadron's insignia – the famous Dutch Girl, trademark of Old Dutch Cleanser. Its significance was explained as "cleaning up on Germany." (Photo courtesy Alan Toelle.)

supplies and information to front-line troops while enduring intense enemy gun fire at very low altitudes. Naturally, numerous losses resulted, as illustrated by the 50th Aero Squadron in its efforts to aid the ill-fated 2nd Battalion of the 308th Infantry Regiment, 77th Division, with whom all contact and hope had been lost. Hence the famous name "Lost Battalion." Yet the battalion was not really lost, but had instead been out-flanked by German forces, while clearing machine gun nests in a ravine near Binarville on October 2, 1918. Existing largely on scant emergency rations for the next five days, the unit remained isolated from its regiment and division, while facing almost certain annihilation by intense German artillery and machine-gun fire that rained down on it.

Meanwhile, flyers of the 50th Aero Squadron made numerous attempts to pinpoint and re-supply the battalion's remnants. Their airplanes flew repeatedly at tree-top level, dropping provisions and messages of encouragement. Despite this, German forces sought the unit's capture and bitterly contested every effort to bring it aid.

On the battalion's fourth day of isolation, October 6, one aircrew from the 50th Aero Squadron – Lieutenants Harold E. Goettler and Erwin R. Bleckley, made two trips over what they hoped was the unit's position. On the first mission, flying less than 100 meters above the ground, their aircraft was seriously damaged by violent gun fire and they were barely able to return to their airfield. Undaunted, the Goettler/Bleckley team borrowed another crew's aircraft and made a second flight hours later. This time they were flying only a few meters above the ground when enemy bullets wounded both men and knocked the plane out of control. In the succeeding crash Goettler was killed instantly and Bleckley so severely wounded that he died before medical aid could be given him.

After the war ended, the United States Army recognized Goettler and Bleckley's act of bravery by awarding each the Medal of Honor posthumously. They

Officers of the First Observation Group during the Argonne offensive. Standing left to right are: A.J. Cole, C.O. 1st Aero Squadron; Winchester, Intelligence Officer; Steward Bird, Chief Observer and Operations Officer 50th Aero Squadron; Stuck, Chief Observer and Operations Officer 12th Aero Squadron; Dan Morse, squadron C.O. 50th Aero Squadron; and Stephen Noyes, C.O. 12th Aero Squadron. (Photo courtesy Dennis Gordon.)

represent the Army's only two-seater crew awarded the Medal in World War One. Appropriately, their citation reads in part that both men had "shown the highest possible contempt of personal danger, devotion to duty, courage and valor."

Erwin R. Bleckley was born in Wichita, Kansas, in the 1890s (the exact date is unknown). He was a bank teller in Wichita when the United States entered the war, but having served with the Kansas National Guard he became active in organizing and recruiting volunteers. His major efforts were towards forming an artillery battery, which later became part of the state's 130th Field Artillery. In late 1917 Bleckley was commissioned a Second Lieutenant for his efforts. Meanwhile, the 130th Field Artillery was attached to the 60th Brigade of the 35th Division then stationed at Fort Sill, Oklahoma.

Coincidentally, the artillery school at Fort Sill was in the process of training observers for the Signal Corps. Upon his arrival, Bleckley found himself attached indefinitely to the Signal Corps and enrolled in the program. In March 1918 he was sent to France to continue his training, first at the First Corps Aerial Observers' School at Amanty, and then at Ourches, where he obtained practical experience with the 90th and 135th Aero Squadrons.

Completing his training in August 1918, Bleckley was attached to the 50th Aero Squadron as a fully trained aerial observer. As such, he was expect to have an ability to accurately fire a machine gun, to "send and receive eight words a minute by telegraph code, to make 12 good aerial

A close-up of De Havilland 4 (serial 32224), "Number 5." The pilot was equipped with two Marlin machine guns. (Photo courtesy Alan Toelle.)

Left: De Havilland "Number 5" was flown by Lieutenant Frayne on October 6, 1918, in his efforts to locate the "Lost Battalion." (Photo courtesy Alan Toelle.)

Below left: Another view of "Number 5." The squadron's identification numbers were in a dark color (probably red) outlined in white. (Photo courtesy Alan Toelle.)

Field, near Fort Worth, Texas, for advanced pilot training, this was difficult to obtain because the United States did not have the necessary aircraft or instructors to teach the more complex aerial maneuvers. As a result, Goettler was sent to France to continue his training in March 1918. When completed in August 1918, Goettler was posted with the 50th Aero Squadron.

The 50th Aero Squadron had been organized at Kelly Field, San Antonio Texas, August 6, 1917, with an enlisted strength of 149 men. The unit sailed on H.M.S *Carmania* for overseas duty on January 9, 1918, landing at Liverpool, England, on January 23. From February 4 to July 3, the squadron was attached to the Royal Air Force for training at Harlaxton Field, near Grantham, England. On July 13, the 50th embarked for France, arriving at Le Havre the following day and St. Maixent on July 17, where the men were issued special equipment – gas masks, steel helmets, and pistols. The unit arrived at the First Corps Aerial Observers' School at Amanty on July 27, and over the next several weeks gathered it personnel and aircraft – American-made De Havilland 4 Liberty Planes.

These machines were flown over from the First Air Depot at Colombey-les-Belles, the first arriving on August 3, 1918. The unit received 18 total, which were divided up into three flights of six planes each.[2] The Flight Commanders were First Lieutenants Goettler, F.T. McCook, and T.H. Hardin.

Meanwhile, the American First Army Air Service was

photographs on 18 assigned locations, to locate and direct artillery fire against enemy batteries, and to conduct a prearranged shoot without error."[1]

Harold Ernest Goettler was born in Chicago, Illinois, on July 21, 1890. He had concluded his studies at the University of Illinois prior to America's declaration of war, but returned after enlisting in the Signal Corps on July 9, 1917, having entered the University's Ground School of Military Aeronautics – a three month course, where he learned under strict military discipline the basic elements of military aviation. Next, Goettler was sent through an eight week primary pilot training program in Canada. He was graduated on January 12, 1918, after he demonstrated an ability to execute simple aerial maneuvers and complete a cross-country flight.

Soon after, Goettler was commissioned a Second Lieutenant and given the right to wear reserve military aviator's wings. Although he was transferred to Taliafero

Two De Havilland 4s lined up, most likely on the field at Amanty, France. The machine in the foreground (serial 32446) is "Number 3." This aircraft was used by the squadron C.O., Daniel Morse, Jr. (Photo courtesy Alan Toelle.)

organized under the command of Colonel William Mitchell on August 10, 1918, and from that date until the first week in September its units and supplies were brought to the front in a effort to support the upcoming offensive at St. Mihiel.

In conjunction with this buildup, the 50th Aero was assigned to the First Corps Observation Group and moved to the aerodrome at Behonne on September 4, and then to Bicquely, just south of Toul, three days later.

On the morning of September 12, when the attack was launched, a combination of low lying clouds, heavy mist, and intermittent rain made aerial observation extremely difficult. Regardless, observation, bombardment, and pursuit squadrons performed their respective missions throughout the daylight hours.

The 50th Aero Squadron had been assigned observation duties for the 90th and 82nd Divisions of the AEF. In this role, Lieutenants Frayne and French had the distinction of being the first of eight crews the unit sent over the front lines that day. Lieutenants Goettler and Bleckley were the second team to take-off that morning, flying the "Number 2" aircraft on a reconnaissance mission from Pont-a-Musson to Montanville. The two maintained an altitude of 500 meters for the one-hour patrol, but due to poor visibility had little to report.

On the second day of operations, September 13, the 90th Division exploited its advances, taking the town of Vilcey-sur-Trey and establishing outposts one half mile south of Villers-sous-Preny. Meanwhile, the 50th Aero

Squadron launched six patrols and experienced its first loss when an observer – Lieutenant Bellows – was killed by ground fire during an infantry contact mission. The plane's pilot – Lieutenant Beebe – escaped unharmed, but the aircraft was seriously damaged and forced to land soon after. Apparently, the Bellows/Beebe team found a break in the clouds and descended to an altitude of only several hundred meters and met their fate. The other crews were unable to find similar holes in the clouds and remained at higher altitudes as illustrated by the crew of the "Number 16" plane – Lieutenants Anderson and Bleckley. The two conducted a two hour flight over Vittonville, Eply, and Mardigny, but due to heavy clouds they maintained an altitude of 900 meters and failed to make contact with the ground troops.

The next day, September 14, was the one day during the offensive marked by favorable weather conditions. Unfortunately, it was also the day marked by a noticeable increase in hostile fighters. Some of the best German fliers operated along the St. Mihiel sector and observation planes were repeatedly attacked. However, in a majority of cases, American observation crews successfully defended themselves and accomplished their missions.

The 50th Aero dispatched nine separate observation flights on September 14. The first mission was flown by Goettler, who took Lieutenant Brill as his observer for a two-hour patrol over Pont-a-Musson and Pagny. Using the "Number 2" plane the crew flew at an altitude of 800 meters and reported enemy troop movements to the 90th

De Havilland 4 (serial 32554), "Number 8," as flown by Lieutenants Phillips and Brown on October 8, 1918. The aircraft as hit and disabled near Binarville, but landed between the lines. (Photo courtesy Alan Toelle.)

A close-up of the Number 8 machine. The squadron's insignia – the Dutch Cleansing Girl, carrying a broom – appeared on each side of the fuselage, the girl facing forward. The girl's dress was blue, with white apron, hood, and stick. The shoes were brown. (Photo courtesy Alan Toelle.)

Division, who had advanced the front line to Villers-sous-Preny and Norry. Hours later, Bleckley departed the aerodrome with Lieutenant Bird in the "Number 14" plane for a two hour flight over Pont-a-Musson and Eply.

On September 15 the 50th Aero flew a record 13 missions. That day, Goettler teamed with Lieutenant Sain in the "Number 2" plane for a late morning infantry contact patrol over Norroy and Villers-sous-Preny. Meanwhile, Bleckley worked with Lieutenant Slater in the "Number 12" plane for an early afternoon flight over Pont-a-Musson and Eply. In both cases the aircraft flew at only 200 meters altitude in an effort to support ground troops. And although the flights were highly successful, the Bleckley/Slater team had a brief encounter with enemy fighters before returning to Allied lines.

The following day, September 16, the American First Army concluded its successful attack on the St. Mihiel salient. In addition to all the assigned objectives being

reached by ground troops in much less than the scheduled time, many unexpected gains were also made on the line east of the Moselle River, allowing the Army to advance to Norroy. Equally impressive were the overall operations of the Air Service, which was able to cooperate successfully with the advancing divisions and with the corps, keeping them informed as to the front's constantly changing position, giving accurate information as to what opposition they would meet, and having the enemy sector constantly under surveillance to prevent German counter-offensives. "In many instances the Air Service aided advancing infantry by strafing enemy troops on the ground, and on several occasions Allied planes forced German soldiers to abandon their posts."[4]

Admittedly, there were several cases in which ground troops did not respond to an aircraft's signal because they mistakenly assumed those machines were not supporting them. Consequently, before the Meuse-Argonne Offensive began, many squadrons invested a considerable effort in ensuring the division or corps units under their care could more easily recognize them. In the case of the 50th Aero Squadron, after being assigned observation duties for the 77th Division, two of the unit's aircraft had the undersurface of their lower left wing "conspicuously" marked with an outline of that division's insignia – the Statue of Liberty.[5]

On September 26, after a three-hour artillery bombardment, the Meuse-Argone offensive was launched. It was marked by the same unfavorable weather encountered during the St. Mihiel offensive. Nevertheless, that morning at 6:00a.m., the 50th launched its first aircraft for a mission over the lines. It was followed by flights at 8:00a.m., 10:30a.m., 12:30p.m., 1:30p.m. and 4:00p.m.

Goettler flew over the front lines twice that day using the "Number 2" plane. On the first mission Bleckley was his observer. However, Lieutenant George filled the observer's cockpit during Goettler's second flight over the lines. Bleckley made a second flight that afternoon too, having teamed with Lieutenant Evans in the "Number 15" plane.

The next day, operations were seriously hampered by very poor visibility, and only four missions were conducted. To make matters worse, two aircraft from the unit were engaged by enemy aircraft, one DH-4 being forced to land at the aerodrome near Duacourt after a combat with six enemy fighters.

On September 28, the DH-4 crewed by Lieutenants McCook and Lockwood was seriously damaged by machine-gun fire from the ground. Moments later the plane crashed into a barbed wire section of "No-Mans Land" (between the German and French lines), the two aviators making good their escape running through a hail of German bullets until they reached French lines.

Over the next two days, the 50th Aero Squadron conducted numerous missions in support of the 77th Division. On September 29, one aircrew realized friendly fire was falling on American troops and put a halt to the barrage. And on September 30, Lieutenants Graham and McCurdy flew their entire mission at treetop level in an effort to locate the front line. These two aviators used the uniforms of the infantry to establish an approximate front, and fired a considerable amount of ammunition into enemy troops as a result.

On October 1, the 50th launched nine missions, three contact patrols and six protection flights for photographic Salmson 2A2s. Admittedly, the protection flights were not as hazardous as the contact patrols, but their success rate was very low because the initial rendezvous took considerable time. To make matters worse, the two aircraft flew at different speeds, making formation flights very difficult. Of the six protection missions undertaken that day, only two were successful.

The 77th Division faired just as poorly that day. Yet their overall situation seemed more serious, especially considering the unit had sustained heavy losses while advancing only a few miles into the Argonne Forest. And now on the evening of October 1, the sixth day of the

De Havilland 4 (serial 32116), "Number 9," shown after a crash-landing, Amanty, France, August 20, 1918. Beyond the squadron's identification numbers carried on the top wing's upper right surface, the aircraft carried a double chevron marking on the top wing's upper left surface. The apex of the chevron pointed inboard, its two bands making an angle of about 60 degrees to the wing's leading and trailing edge. The band width of each chevron appears to equal the width of an outer ring on the wing roundels. The space between the two chevrons appears to equal half the width of one chevron band. The inner band was a very light color (probably a light blue). The outer band was a dark color (probably red). (Photo courtesy Alan Toelle.)

offensive, General Robert Alexander, commander of the 77th Division, drew up yet another plan to crack the Argonne. This featured a three-pronged drive from the south, with the center prong of the attack being the 2nd Battalion of the 308th Infantry under the command of Lieutenant Colonel Charles Whittlesey.

Whittesey considered his orders with growing gloom and quickly protested that not only was his battalion at half strength, but the remaining men were exhausted from nearly a week of fighting. Nevertheless, the next morning, October 2, the battalion was ordered to proceed. And with

Lieutenant Colonel Charles Whittlesey, Commander of the "Lost battalion." (Photo courtesy National Archives.)

The carrier pigeon *Cher Ami* (Dear Friend) was the battalion's last pigeon. His flight from the Charlevaux Valley on October 5, 1918, to the 77th Division's pigeon loft at Rampont took about 15 minutes. The bird's leg had been shattered, one wing badly injured, and its breastbone was broken by shell fire, but the message was there, dangling from the capsule on its leg. (Photo courtesy National Archives.)

little choice, Whittesey advanced as casualties mounted at a frightening pace. Within hours Whittlesey also realized his flanking units had been pushed back, leaving his battalion in a dangerous position. In response, he contacted headquarters to request a halt before being cut-off by German forces. To his surprise, General Alexander ordered him to continue forward. For a short time he did, however, as the battalion entered an elongated ravine, he halted its advance for the night, with the hope flanking units would rejoin the battalion in the morning.

Instead, during the early hours of October 3, the Americans found themselves completely surrounded by German forces, and except for the battalion's six carrier pigeons, totally cut-off from all forms of communication.

Whittlesey dispatched three pigeons that day to request immediate aid. The last bird carried a message which indicated the unit's effective strength had been reduced from 600 men to only 245. The note also provided coordinates for the battalion's position, but unfortunately these were wrong, and caused the 50th Aero Squadron to look elsewhere for the besieged unit. To make matters

worse, when aircraft did fly over their position, the soldiers were reluctant to display recognition panels for fear enemy sharpshooters would see their movements first.

In an effort to improve the situation, Wittelsey dispatched two more carrier pigeons on October 4. Besides sending only slightly improved coordinates, one message stated in part: "Men are suffering from hunger and exposure, and the wounded are in very bad condition. Cannot support be sent at once." Sadly, the only contact that day came from the Germans, who had offered the battalion an invitation to surrender. Admittedly, by this point the Americans were existing largely on scant emergency rations, and facing a growing possibility of annihilation as the battalion fighting strength continued to shrink. Nevertheless, Whittlesey refused the offer with a scornful reply: "Go To Hell."

De Havilland 4 serial 32169 "Number 2" as flown by Lieutenants Erwin Bleckley and Harold Goettler on October 6, 1918. This aircraft was finished in a two-color pattern typical of Dayton-Wright DH-4s: the upper wing and fuselage surfaces were finished in khaki, and yellowish-gray was applied to the remainder of the fuselage, lower wing surfaces, wheel covers, and struts. The "2" appears under the lower starboard wing, on the fuselage, and on the upper wing. (Photo courtesy R.L. Cavanagh.)

The next morning, October 5, the 77th Division made an attempt to break through to the battalion, but within hours they were repulsed. The division's artillery units reacted next, laying down a four hour barrage on what should have been German positions, if the coordinates Whittlesey provided were correct. Instead, the barrage came down upon Whittlesey and his men, who were at first elated to hear guns firing from the American lines, but ended up scrambling for cover once shells started falling.

During the attack, Whitlesey disregarded his own safety by running to the pigeon handler and attaching a note to the battalion's last bird. He then tossed the pigeon into the air only to watched it fly to a nearby tree and perch. In desperation, Whittlesey persuaded the bird to fly through the artillery barrage by throwing stones at it.[6]

About 15 minutes later, the pigeon, by then badly injured, reached division headquarters with the following note: "We are along the road parallel 276.4, our artillery is dropping a barrage directly on us. For heaven's sake stop it."

The artillery barrage was immediately stopped, but another 80 men had been killed or wounded. Hope of reaching the battalion in time was believed "lost." However, the coordinates in Whittlesey's last message "along the road paralleled 276.4," when combined with others received from previous messages, indicated a position of "294.7–276.3." These coordinates were very close to the battalion's actual location. And when the aviators of the 50th Aero Squadron were called upon to drop food, ammunition, and medical supplies on October

6, some provisions actually reached the battalion. However, with 13 separate missions flying over the same general area, it became apparent to the Germans that a determined effort was being made that day. Consequently, with the arrival of each succeeding plane, enemy gun fire exacted a toll.

The first mission reached the drop zone slightly before noon, with Lieutenants Pickrell and George flying the "Number 6" plane. It was quickly followed by several others, and fortunately these crews escaped without serious trouble; however, the situation quickly deteriorated.[7] Lieutenant Morse returned with his "Number 3" machine having nearly 40 bullet holes in it, as did the "Number 2" plane flown by Goettler and Bleckley. Next to be hit by machine-gun and rifle fire was the "Number 8" aircraft flown by Lieutenants Phillips and Brown, which crash-landed northeast of Binarville soon after. The "Number 14" plane endured a similar fate and crashed near Vienne-le-Chateau with Lieutenants Bird and Bolt on board. In both cases these men escaped serious injury and made it back to the squadron. Not as fortunate was Lieutenant Slater, who took a beating in the "Number 12" plane, being seriously wounded in the foot and barely able to return to base.

Meanwhile, Bleckley and Goettler believed they had sighted the Battalion during their first flight and so volunteered for a second. But since the "Number 2" aircraft had been so badly damaged by enemy fire, the two men set out in Lieutenant Pickrell's "Number 6" plane that afternoon, but never returned.

The Goettler/Bleckley "Number 2" machine. At the time of this photo the squadron's identification number had been applied to the fuselage. However, the unit's insignia – the Dutch Cleansing Girl, carrying a broom – had not. (Photo courtesy Dennis Gordon.)

The last crew to go out that day, Lieutenants Graham and McCurdy in the "Number 1" plane, also ran into trouble. They had descended to treetop level, and McCurdy was preparing to drop a basket of carrier pigeons, when several Germans came out of a dug-out and fired a bullet through McCurdy's neck. Although this was a serious, painful wound, by his grit and nerve, McCurdy gave a full report while his wounds were being dressed at the field hospital.

Out of desperation, the next morning, October 7, Lieutenants Anderson and Rogers took the "Number 16" plane over the area and proceeded to circle continually at a low altitude until they spied a recognition panel and charted the battalion's exact position. With their precise location known, the Americans made another drive towards Whittlesey and his men. By day's end they had succeeded in rescuing the survivors.[8]

Coincidentally, that same day, the 50th Aero Squadron learned that Goettler and Bleckley had been killed during their second patrol on October 6. The two had been flying only a few meters above the ground when enemy bullets wounded both men and knocked the plane out of control. In the ensuing crash Goettler was killed instantly and Bleckley so severely wounded that he died before medical aid could be given him. The report from the French troops was that both men were buried.

In 1922 the two men were awarded the Medal of Honor posthumously. Bleckley's citation reads: "2d Lt. Bleckley, with his pilot, 1st Lt. Harold E. Goettler, Air Service, left the airdrome late in the afternoon on their second trip to drop supplies to a battalion of the 77th Division, which had been cut off by the enemy in the Argonne Forest. Having been subjected on the first trip to violent fire from the enemy, they attempted on the second trip to come still lower in order to get the packages even more precisely on the designated spot. In the course of his mission the plane was brought down by enemy rifle and machine-gun fire from the ground, resulting in fatal wounds to 2d Lt. Bleckley, who died before he could be taken to a hospital. In attempting, and performing this mission 2d Lt. Bleckley showed the highest possible contempt of personal danger, devotion to duty, courage, and valor."

Goettler's Medal of Citation reads the same as Bleckley's.

Charles Hazeltine Hammann

Landsman for Quartermaster Charles Hazeltine Hammann. His awards included: U.S. Medal of Honor; Italian *Al Valore Miltare* (Medal for Valor), in silver; and the Italian *Croce di Guerra* (War Cross). His Italian pilots badge is clearly displayed on his upper left breast. (Photo courtesy Dennis Gordon.)

Charles Hazeltine Hammann was born in Baltimore, Maryland, on March 16, 1892. By the time he graduated from Baltimore's Polytechnic Institute, Hammann had – to a small degree – a basic understanding of flying. Therefore, when enlisting on October 27, 1917, he was assigned to the Navy's First Aeronautical Detachment with the rank of Landsman for Quartermaster.[3] With this unit, Hammann served aboard the U.S.S. *Neptune* at Baltimore, Maryland, and later at Norfolk, Virginia. He sailed for France aboard the U.S.S *Neptune*, arriving at St. Nazaire, France, on June 8, 1918. In rapid succession, Hammann rotated through the Tours and Moutchic flight training centers. Then he entered the French Naval School of Aerial Gunnery at Lake Hourtin. In early July 1918, he was sent to Italy, where he would finish his pilot training.

Months earlier, on November 21, 1917, the Italian government had requested that the U.S. Navy extend its aviation activities into the Adriatic Sea. In support of this plan, the Italians offered to turn over three established air stations to the Americans – San Severo, Pescara, and Porto Corsini – and fully equip each, if the U.S. Navy would man and operate them.

Following a detailed evaluation of the Italian installations, the U.S. Navy agreed to take over two – Pescara and Porto Corsini – but also indicated a desire to obtain control of the Italian Naval Air Training Station at Lake Bolsena, located about 60 miles north-east of Rome.[4]

The other two stations – Pescara and Porto Corsini – were strategically situated on the Italian coast of the Adriatic Sea. Both were less than 70 miles from the Austrian naval base at Pola, a critical target to the Allies because it was common for battleships and cruisers of the Austro-Hungarian High Sea Fleet to be anchored there. Furthermore, German and Austrian submarines used Pola as a staging port in their campaign to control the Mediterranean Sea.

The Allies envisioned U.S. Naval Aviators from Pescara and Porto Corsini would rendezvous with Italian aviators from Venice, some 50 miles to the north, for combined aerial attacks on the Austrian naval base at Pola.

The Austrians also foresaw aerial attacks on Pola and established a network of 18 forts and heavy gun batteries to defend the naval base and city from Allied bombing. In all, an estimated 115 anti-aircraft guns were in place, making Pola a formidable target for any attacking force.

The Americans acknowledged this by admitting extensive training would be required before any Navy mission was launched against this target. In conjunction with that, on February 19, 1918, the first American aviation detachment to enter Italy arrived at Bolsena. Two days later, command of that station was officially transferred to the U.S. Navy.

Actual instruction at Bolsena started immediately. The first flying instructors available were Italian, but they were replaced with American personnel once the Americans

Charles Hazeltine Hammann has the distinction of being the only U.S. Naval Aviator awarded the Medal of Honor for action against enemy forces during the First World War.[1] Yet, if one considers Hammann's pugnacious nature and the fact he was stationed at Porto Corsini, Italy, it should not be surprising he was highly decorated. In fact, so aggressively did that installation carry out its mission against enemy forces that Admiral H.T. Mayo, USN, stated on the basis of his inspection of November 10, 1918, that the Porto Corsini naval base had "the distinction of being the most heavily engaged unit of the U.S. Naval Forces in Europe." In truth, the squadron stationed at Porto Corsini was not only the most heavily engaged U.S. naval aviation unit in World War I, it was also the most decorated. Of the 30 U.S. Navy pilots who flew missions against Austria-Hungary's main naval bases of Pola and Trieste on the upper Adriatic coast, 14 were awarded Navy Crosses and one the Medal of Honor.[2]

became proficient at handling the Italian aircraft. Initially, F.B.A. flying boats were used, but these planes were exchanged for Macchi L-3, M-8, and M-5 flying boats. To provide maintenance for these planes, "a special draft of mechanics was assembled from the men training at the various Italian seaplane and motor factories."[5]

The Bolsena Ground School curriculum eventually consisted of courses in theory of flight, navigation, engines, signaling, standard navy regulations, and the Blue Jacket's Manual.

In a letter from Lieutenant (jg) Walton (a student at Bolsena), the following flight training philosophy was given: "A total of 12 hours of flight instruction was provided; if the student did not solo, two more flights of 30 minutes each were made; if it were felt solo would occur within 1–2 more hours, training was continued; if not, all flying was stopped."[6]

In total 134 American aviators underwent training at Bolsena, many filling the ranks at Porto Corsini and thereby allowing the U.S. Navy to officially transition into that station. Hammann arrived at Bolsena in early July 1918. Upon completing the flight training course he was awarded his wings (U.S. Naval Aviator number 1494) and assigned to the naval base at Porto Corsini on July 24.

Coincidentally, on that same day Lieutenant Commander Willis Bradley Haviland, USN, took control of Port Corsini and placed it in commission as a U.S. Naval Air Station. This was a remarkable feat, considering Haviland and his aviation personnel – a detachment of 35 officers and 330 enlisted men – had arrived at this base from Paullic, France, only the day before. Nevertheless, the U.S. Navy commenced operations immediately and quickly Americanized their surroundings – unofficially naming the base "Goat Island" after the U.S. Navy mascot.

From an aerial view, the U.S. Navy Air Station at Porto Corsini appeared to be a small V-shaped island, with two narrow, 100 foot wide canals straddling the base on either side and meeting at a junction not far inland. On closer inspection, the aviation facilities at the base were of ample size, consisting of a long series of well fabricated hangars situated along the inside junction of the two canals. The area surrounding the island base was generally marsh-swamp, infested with malaria-carrying mosquitoes. Yet the sanitary precautions were so thoroughly enforced at the station that the health of the personnel was always excellent.

Beyond this issue, the station had two tremendous disadvantages. First, there were never enough aircraft; only three flying boats were made available to the Americans in the initial stages of operating the base. And although the number of planes quickly increased, there were never more than 18 altogether.[7] Secondly, the small flying boats used at Porto Corsini were not well suited to open sea conditions; therefore all landings and take-offs had to be made on one of the narrow canals. This, combined with the orientation of the prevailing winds – normally 90 degrees to the direction of the canal – made for a real handicap when pilots tried to take-off and land. The second disadvantage was counteracted to some extent at Lake Bolsena, by training the student pilots to land on an

A Macchi M-5 flying boat landing in one of the narrow canals that straddled Naval Air Station (NAS), Porto Corsini, July 1918. (Photo courtesy Noel Shirley.)

Macchi M-5 flying boats on the concrete dock at NAS Porto Corsini, circa July 1918. (Photo courtesy Noel Shirley.)

area, marked off by buoys, which equaled the 100 foot width of Porto Corsini's canals. Students practiced under these demanding conditions in the Italian flying boats until they were proficient. Yet it was perhaps to Haviland's credit that, in spite of the extreme difficulty his pilots faced in departing from and landing on Porto Corsini's narrow, wind-swept canals, the squadron's fatalities were limited to only four personnel.

The Austro-Hungarians were aware the Americans had arrived at Porto Corsini on July 24, and welcomed them by launching a bombing attack on the following night. Fortunately for the Americans, the Austrians miscalculated the location of the base and a majority of bombs landed in the swamps and canals further up the coast.

Despite the ineffectiveness of the Austrian attack, the U.S. Navy pilots at Porto Corsini were quite eager to retaliate. However, with overall operations under the control of the Italian military, the Americans were held back from bombing missions and their activities were "devoted mainly to the dropping of propaganda leaflets."[8]

The Americans did not understand the importance of these missions. They were, in the Americans' opinion, non-combat patrols, and therefore unacceptable. In fact, dropping of propaganda material was perhaps the most dangerous method of warfare on the Italian Front, because the "Austrians had retaliated by threatening to shoot as spies all fliers engaged in such duty."[9]

Ansaldo S.V.A. 5 floatplane. Italy adapted its famous S.V.A. fighter to floats. It was initially envisioned that Porto Corsini would receive six Ansaldo S.V.A. 5 floatplanes for defending the station against air attacks. However, it is believed none ever reached the station.

This detail did little to dampen the aggressiveness of the Americans at Porto Corsini. Haviland had hand-picked his men, enlisting when possible former Lafayette Flying Corps pilots, the U.S. volunteers from the French Air Service who transferred to the U.S. Naval Air Service, and who Haviland knew would carry out his motto: "Go get 'em – Talk about it afterwards."[10] In fact, several weeks later, on the morning of August 21, American ingenuity found the excuse and the means to hit Pola.

That day at 10:30 a.m., a patrol of seven aircraft departed Porto Corsini with orders to drop propaganda leaflets on the base at Pola. During the first 20 minutes of the mission, two aircraft had to abort because of engine problems, a Macchi M-8 (two-seat flying boat), piloted by Lieutenant Read with Lieutenant Haviland as observer, and a Macchi M-5 – serial 7293 (single-seat flying boat), flown by Ensign Elmer Johanson. The remaining aircraft which continued on were: a Macchi M-8 (serial 19008) piloted by Ensign Walter White with Ensign Albert Taliaferro as observer, and four Macchi M-5s – serial 13015, flown by Ensign George H. Ludlow, the flight leader; serial 7299, flown by Ensign Austin Parker; serial 7225, flown by Ensign Dudley Voorhees; and serial 7229, flown by Landsman for Quartermaster Charles Hammann.

The American patrol approached Pola from the south at 11:20 a.m., the slower M-8 at 2500 meters altitude and the M-5s at 3800 meters. Wasting little time, the group unloaded their propaganda leaflets and then established a westerly flight path – towards Italy – while enemy anti-aircraft guns were putting up heavy but poorly directed fire. At the same time, some of the Americans noticed "five *chasse* [sic] planes and two hydroplanes" taking off and rising rapidly towards them. However, it is likely that the two seaplanes were incidental traffic over this busy naval base rather than slow-flying interceptors that never caught up with the Americans flight – as "nothing further was seen of the hydroplanes." In any event, the enemy land-chase planes were attempting to intercept, and the American pilots would soon have their hands full.

Unfortunately for the Americans, just prior to this mission the Austro-Hungarians had formed a specialized fighter detachment to defend Pola from steadily increasing air attacks. This unit was stationed at Altura airfield, just inland from Pola, and equipped with the Phönix D.I land-based fighter – a fast, dependable, rugged aircraft.

When the Americans appeared over Pola, the fighters at Altura airfield were called to intercept them. The four pilots who responded were *Fliegermaat* (sergeant) Josef Gindl, flying aircraft A-102, *Fregattenleutnant* (naval lieutenant) Emil Prambergr in A-111; *Fregattenleutnant* Stephan Wollemann in A-118; and, piloting A-117, *Fregattenleutnant* Friedrich Lang, the Austro-Hungarian Navy's only ace besides Baron Gottfried Banfield.[11]

Believing that the enemy fighters would soon engage

U.S. Naval Officers, Porto Corsini, August 10, 1918. Seated (left to right): G. W. Knowles, E. I. Tinkham, C. H. Hammann, J.A. Goggins, J. Stanley. Standing (left to right): R. H. Clark, E. L. Smith, C. W. Gates, W. White, A. P. Taliaferro, K. Stewart. Note the diverse style of pilots 'wings' worn, including Italian and French designs. (Photo courtesy U.S. Naval Historical Center.)

White and Taliaferro in the slower Macchi M-8, Ludlow gave the signal for the M-5s to dive down onto the enemy aircraft. However, Voorhees could not follow; when testing his guns beforehand, the springs in his cartridge clips broke, jamming both weapons hopelessly. He was forced to remain above the battle as a frustrated observer. Despite this, Parker and Hammann quickly followed Ludlow into the fight, and at 11:25 a.m., fliers of Altura and Porto Corsini met just west of Pola at an altitude of 2500 meters.

Still diving, Ludlow initially fired at the lead Phönix, forcing it into an evasive dive. Parker followed this Phönix down while firing both machine guns, but his aircraft's right gun jammed before he inflicted serious damage. At this point Parker pulled up and, after emptying his left gun at two of the enemy aircraft above him, retired and attempted to clear his right gun.

Meanwhile, Ludlow engaged a second Phönix while Hammann battled two others. At this point the remaining participants entered what is best described as a 'melee.' Each of the Macchi flying boats and Phönix fighters made multiple attacks on one another in the ensuing battle, but in no particular pattern. The quarters were very close and

the gunnery accurate, as evidenced by the fact that every aircraft involved suffered damage.

Early in the battle, Ludlow's gun fire hit the Phönix fighter piloted by *Fliegermaat* Gindl; it fell away trailing white smoke and was not seen again by the Americans. Ludlow felt he had "apparently perforated the radiator" of this machine. Voorhees supported this claim when he later reported "seeing the enemy machine dive away trailing smoke." To this day, Ludlow is credited with having put that Phönix fighter out of action by shooting up its engine. However, this was not the case. Soon after opening fire, the phosphorous ammunition loaded in *Fliegermaat* Gindl's Phönix D.I had suffered from spontaneous combustion. He retired from the battle immediately, making a successful forced landing near the anti-aircraft batteries at Valbandon.

Meanwhile, Ludlow was having serious difficulties of his own. His Macchi had suffered multiple hits and he attempted to fly away from the fight. Unfortunately for him, two enemy fighters followed. *Fregattenleutnant* Wolleman in A-118 was one of Ludlow's pursuers, and he hit Ludlow's Macchi with repeated bursts of machine-gun

Macchi M-3 flying boat. The M-3 was an all-purpose aircraft, used for reconnaissance, bombing, patrol, and combat. Power was supplied by a 160 hp Isotta-Fraschini IF-V-4B engine. Many of the U.S. Navy officers assigned to the Naval Training Station at Bolsena, Italy, learned to fly on the M-3.

Macchi M-8 flying boat. The M-8 first appeared in late 1917. It had a top speed of 102 mph and could accommodate a crew of two or three, depending on its role. It was typically equipped with a single Fiat-built Revelli machine gun, which was mounted on a rotating ring located in the bow. NAS Porto Corsini was equipped with several M-8s.

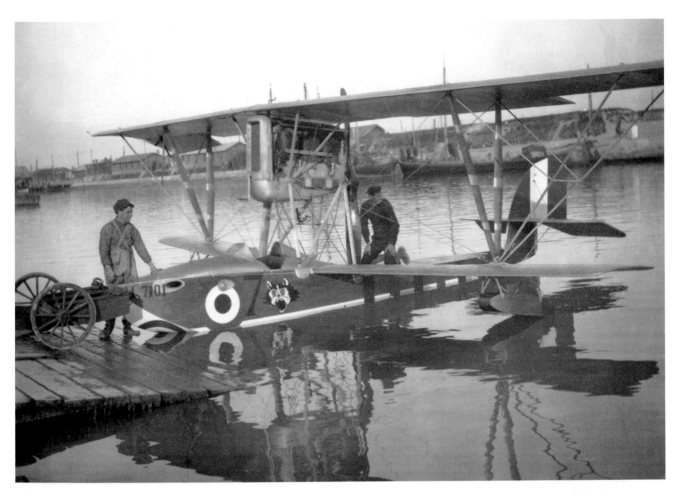

Macchi M-5, serial 7101. The black bands on the hull's sides indicate this machine was assigned to Italian naval squadron 252A, which operated out of Venice, Italy. The last bombing mission involving airmen from Porto Corsini occurred on October 22, 1918. For this attack on Pola, 30 aircraft from Venice rendezvous with the Americans.

A close-up of an F.B.A. flying boat. The large number of bombs carried by these machines is illustrated clearly.

A close-up of a captured Macchi M-5, serial 1354. The interplane bracing consisted of tape-bound wooden V-struts, with wire-braced tubular metal struts supporting the engine.

Ensign George H. Ludlow. Hammann received the Medal of Honor for saving his life. (Photo courtesy Dennis Gordon.)

A close-up of a Macchi M-5's 160 hp Isotta-Fraschini V4B engine and its pusher propeller, Porto Corsini, circa 1918. The pilot is sitting just below the radiator.

Macchi M-5 flying boat, serial 4870. Even on a beaching trolley this aircraft lost none of its elegance. The M-5 was a small single-seat fighter which first appeared in early 1918. The wings were constructed of wood and wire braced. Once assembled, wings were fabric covered and coated with a protective varnish. The plywood hull was also covered with a protective varnish. But the bottom surface of the boat hull and sides up to the water line were typically painted with a marine point – usually white. The M-5 was fully aerobatic and had a top speed of 118 mph.

Macchi M-5, serial 13015, flown by Ensign George H. Ludlow on August 21, 1918. The top section of the bow (front) has been painted black to the cockpit and along the sides of the bow to a point just beyond each roundel. The bottom surface of the hull was painted with a white marine paint to a point reaching the lower bow roundel. This same white paint was applied on both sides of the varnished wooden hull in narrow bands; creating alternating bands of white paint and varnished wood. The outer and inner sides of the wing tip floats were also painted white. Ludlow's personal insignia appeared in red on at least the starboard side of the Hull – *MUTT 2nd.* Wing undersurfaces were in standard Italian tri-colors. Roundels had red centers and green outer rings.

fire, shooting off its right magneto and striking its propeller and, more seriously, shooting a hole in the engine's crankcase, causing it to lose its oil. The engine of Ludlow's Macchi soon burst into flames. In a desperate attempt to shake the Austro-Hungarian fighters, Ludlow cut off power and dove his flying boat steeply toward the water. Apparently, at the sight of Ludlow's fighter in near-vertical descent, trailing flames from its engine, Wollemann and his wingman felt the American was finished and at 500 meters altitude broke off their attack and returned to Altura.

The severity of Ludlow's maneuver not only convinced the Austro-Hungarians to break off pursuit, but it also put out the fire in his engine. Maneuvering skillfully with a dead engine, Ludlow managed an adequate landing on the water, three miles off Pola. Hammann had seen Ludlow go down and followed him. Arriving over the downed Macchi, he courageously attempted a rescue. As Hammann landed and taxied up to the damaged Macchi, Ludlow was working to ensure that his flying boat would not become an enemy trophy. He opened the photographic port in its hull, allowing it to fill with water. He also kicked several holes through the wooden hull and slashed the fabric on the wings before swimming over to Hammann's craft.

Unfortunately, the cockpit would not accommodate both men, so Ludlow crawled in behind Hammann – under the engine – and laid down flat on his stomach, hanging onto the struts of the engine mount. But due to water entering through bullet holes sustained in the battle, the added weight of Ludlow, and the three-foot choppy seas, Hammann had considerable trouble taking off and seriously damaged the nose of the flying boat's hull before finally becoming airborne.

After firing 100 rounds of ammunition at Ludlow's fighter to ensure it would sink, Hammann and Ludlow returned to Porto Corsini, over 60 miles away. Hammann made a perfect landing in the canal, but the damaged nose of the Macchi took on water immediately. After traveling for only a few yards across the canal, the fighter nosed down and flipped over, trapping Hammann by his safety belt. Ludlow dove under and freed him. Both pilots were quickly rescued by naval personnel. And except for minor injuries – Ludlow with a gash on his forehead, and Hammann with a strained side and shoulder – neither had been seriously hurt.

After hearing details of the mission, the commanding officer of the U.S. Naval air stations in Italy, Lieutenant Commander J.L. Callan, composed a recommendation for Hammann and Ludlow. It read in part:

"The work done by Hammann by going to the aid of Ludlow immediately, was a fine exhibition of courage and head-work on his part, and his having been able to get off the water with a single-seat chasse machine with two people aboard and bring back Ludlow safely, was an excellent performance, and he is deserving of a great deal of credit for his quick action and good judgment. He undoubtedly saved Ludlow from being taken prisoner, and his machine at the time of landing was only 5 miles [*sic*] from Pola harbor.

"At the time he did this, he knew that there was a

Fregattenleutnant Friedrich Lang of the Austro-Hungarian Naval Air Service, circa 1916. Lang, the Austro-Hungarian Navy's only ace besides Baron Gottfried Banfield, ended the war with five confirmed victories.

possibility of his being attacked by the Austrian machines when he was on the water and this case is undoubtedly one which shows a great deal of courage and bravery. It is therefore strongly recommended that Hammann, Charles H., Landsman for Quartermaster, U.S.N., be given the Distinguished Service Cross of the U.S. Government, if it is possible to present this to the Naval Aviators, or with some other decoration for his very creditable act. It is also recommended that Ensign George H. Ludlow, U.S.N.R.F. be commended for his action in putting an enemy machine out of the fighting by damaging its motor and causing it to return to its base."

That night the Austro-Hungarians retaliated against Porto Corsini, dropping 49 bombs from ten planes. Although one building on the base was hit, there were no casualties. The remaining bombs fell harmlessly across the canal.

Not allowing this enemy attack to go unchallenged, the next evening, Ensign R.B. Read and Quartermaster George

Above: *Fregattenleutnant* Stephan Wollemann, seated on the wing of his Phönix D.I fighter at Altura, shot down Ensign Ludlow for his only victory in the war.

Below: Phönix D.I (A98) at Altura, winter 1917–18.

Lieutenant Commander Willis B. Haviland, standing by one of his squadron's Italian built Macchi M-5 flying boats, No.31. The hull and wing struts of this particular machine have been painted in alternating bands of black and white. Further decorations include a small skull and cross-bones and navy goat insignia. Considering U.S. Navy personnel at Porto Corsini unofficially named the base 'Goat Island' after the U.S. Navy mascot, it is no wonder this insignia appeared on several aircraft. The gray goat had a yellow blanket, and the word *NAVY* appeared in white. (Photo courtesy Dennis Gordon.)

M. Knowles left Porto Corsini in two Macchi M-8s. They scored hits on Pola's defenses and aircraft hangars. At this point a pattern of being bombed and retaliating had been established. The Americans launched additional attacks at Pola on August 24 and 29, dropping a considerable number of bombs on the enemy base.

On October 7, six Macchi M-5s left Porto Corsini at 7:00 a.m. for a reconnaissance patrol, reaching Pola at 8:20 a.m. The Macchi M-5's were subjected to anti-aircraft fire and five enemy aircraft took to the air; however, they did not engage the American machines as they made two complete circles of the harbor before returning safely to Porto Corsini at 9:00 a.m.

The last bombing mission involving airmen from Porto Corsini occurred on October 22; eight Macchi M-5s, three Macchi M-8s, and two F.B.As rendezvous with 30 naval aircraft from the Italian squadron at Venice for a combined aerial attack against Pola. The barrage put up by the Austrians was very poor and did little to divert the Allied planes in their bombing run on the enemy station.

Before the end of October, the bombing stopped in Italy. Shortly after, in November 1918, the Armistice between Italy and Austria-Hungary took effect. Following the cessation of hostilities, the Italian Government decorated 17 of the station's pilots with medals of valor. Hammann was awarded the Italian *Al Valore Miltare* (Medal for Valor), in silver; and the Italian *Croce di Guerra* (War Cross).

Hammann had been appointed to the provisional rank of Ensign in the U.S. Naval Reserve Force (Flying Corps) on

Charles H Hammann standing in front of a Macchi M-8 flying boat, No.34, serial 13085, NAS Porto Corsini, circa November 1918. Portions of the hull have been painted black. Further decorations include the Navy goat insignia.

October 14, 1918. He remained in the European Theater until January 21, 1919, when he returned to the United States for duty at the U.S. Naval Station at Bay Shore, New Jersey. Soon after, Hammann was posted with the Naval Air Detachment at Langley Field, Virginia. On June 14, he was representing the Navy at a flying event at Langley Field when, without warning, his Macchi built M-5 flying boat fell into a spin and crashed on land at 1:50 p.m. Hammann was pinned under the wreckage and believed to have died immediately by injuries sustained in the crash. At the time of his death, Hammann was about to be transferred to the Regular Navy, with orders to report to the U.S.S. *Oklahoma* (BB-37) for duty as an aviator.

A court of inquiry noted at the time of the accident Hammann was flying level at an altitude of 1200 feet and had not performed any previous stunting or aerobatics in the Macchi M-5. The court attributed the accident to either a break in a control cable fitting or to the probability that this type of aircraft would not recover from a spin on dead center due to the position of its engine.

Hammann was buried at Oak Lawn Cemetery in Baltimore, Maryland. Weeks later, on July 9, 1919, Commander J.L. Callan USNRF, drafted a letter to Rear Admiral Sim's in which he "strongly recommended" that Hammann be posthumously awarded the Medal of Honor. Callan stated that Hammann's action in the rescue of Ludlow "was an extraordinary exhibition of self sacrifice as Hammann knew that if he failed, there was no chance of himself being rescued and that both would be taken prisoner by the Austrians. His act was exceptionally meritorious and commendable."

Hammann was posthumously awarded the Medal of Honor. To pay further homage to his courage, the United States Navy named a destroyer, the U.S.S. *Hammann* (DD-412), in his honor. His ship was launched on February 4, 1939. After DD-412 was sunk in the Battle of Midway in 1942, a destroyer escort was named U.S.S. *Hammann* (DE-131) and commissioned in the U.S. Navy on May 7, 1943.

Frank Luke, Jr.

Frank Luke, Jr. His military awards included the Medal of Honor, Distinguished Service Cross with Oak Leaf Cluster, and the Italian *Croce di Guerra*. (Photo courtesy NASM.)

Frank Luke's contemporaries described him as a physically strong, clean-looking young man. With blond hair and piercing steel-blue eyes, he possessed mannerisms that always conveyed quick, energetic motions. In the assessment of Luke, many would also add he was very self-confident, extremely daring, highly individualistic, and totally antagonistic to any discipline whatsoever. Perhaps more than anything else, two of Luke's personal traits – overconfidence and brashness – made a poor impression on people and quickly earned him a reputation as a braggart and malcontent, to the point his friends were almost non-existent. Addressing this issue, Major Hartney, commander of the First Pursuit Group, agreed that Luke's remarks had certainly conveyed a high degree of boastfulness. Nevertheless, he personally believed Luke's comments were nothing of the sort, suggesting instead they "were simply the honest confidence of a zealous but not-too-diplomatic boy."[1] Unfortunately, many individuals interpreted Luke's actions less favorably, and as a result, Hartney indicated they "harshly preyed on his mind to such an extent that he became almost a recluse with an air of sullenness."[2]

In the end, many continued to sharply criticize Luke's method of conducting business in spite of his phenomenal success during a three week period in September 1918. However, few could contest Hartney's statement in which he regarded Luke as "the most extraordinary flier ever produced by the United States Army, with an official record of achievement at the front which was both breathtaking and unequaled."[3] Luke's achievements, based largely on his extremely daring nature, made him the second highest scoring American 'ace' of the war and earned him the Medal of Honor posthumously.

Frank Luke, Jr. was born May 19, 1897 in Phoenix, Arizona, one of nine children born to German immigrant parents. He was educated at Phoenix High School, where he excelled in baseball, football, and track. In September 1917 Luke enlisted in the Signal Corps at Tucson, Arizona and was assigned to the Ground School of Military Aeronautics in Austin, Texas. Graduated in November, he next went to the flight training center at Rockwell airfield, located at San Diego, California.[4] Luke was commissioned a Second Lieutenant in January 1918. Then in March he sailed from New York for France aboard the liner *Leviathan* (ex-*Vaterland*).

In France Luke obtained additional training at the Third Aviation Instruction Center at Issoudun. There he first encountered a troubled young man, Joseph Wehner, who later became his closest friend at the front and who saved his life on more than one occasion.

Joseph Fritz Wehner was born in Boston, Massachusetts on September 20, 1895. Wehner's parents, like Luke's, were both German immigrants. His father, Fritz Wehner, was a cobbler in Boston.

Ironically, Wehner's own troubles resulted from humanitarian activities after he sailed to Germany, where he worked with the YMCA distributing care packages to Allied prisoners of war (POWs). Although Wehner immediately left Germany and returned to the United States after the two countries declared war, he was suspected of anti-American activities due to his Teutonic ancestry, and temporarily held by Secret Service officials. Wehner was released and allowed to enlist in the air service. However, he was under constant observation at Scott Field in Belleville, Illinois, and Kelly Field, Texas, where he conducted his aviation training. At Kelly Field, Secret Service officials placed Wehner under temporary arrest, but he was duly acquitted and received his commission as a First Lieutenant on January 9, 1918. Wehner sailed for Europe as part of the AEF. While training at Issoudun, France, he was questioned by government agents again. Although no official action came from his

Luke in flying leather coat and helmet. (Photo courtesy NASM.)

Lt. Wehner photographed while training at Issoudun, France, circa spring 1918.

interrogation, the damage was done – Wehner had become withdrawn and isolated.

Meanwhile, Luke had completed his training at Issoudun and was assigned to the American Aviation Acceptance Park at Orly, ferrying new aircraft to front-line squadrons. In July 1918, Luke was transferred to the First Pursuit Group, which in turn placed him with the 27th Aero Squadron, then located in the Aisne-Marne Salient.

The 27th Aero Squadron had been officially organized on June 23, 1917, from Company "K" of the Third Provisional Aero Squadron, located at Camp Kelly, later Kelly Field, San Antonio, Texas.

After a short training program in Canada, the 27th Aero was transferred to Taliaferro Field, near Fort Worth, Texas, under the command of Major H.E. Hartney. The squadron left Texas in January 1918, moving to Mineola Field in Garden City, New York. Subsequent to the usual processing and waiting, the squadron boarded ship for travel overseas, sailing on February 26, 1918. The 27th Aero reached England on March 5 and departed five days later for France. The squadron initially trained at Tours, France, but concluded its training at the Issoudun facility.

On June 1, the 27th Aero Squadron moved to the aerodrome at Toul as part of the American First Pursuit Group.[5] The next day, June 2, active operations for the squadron began at 6:00 a.m. when the unit's first patrol took-off for a view of the front lines.

Gradually the 27th Aero gained experience and began to actively scout the lines from St. Mihiel to Pont-a-Mousson. Although aerial activity was comparatively low for the first week, the squadron's first combat occurred during the mid-day patrol of June 6 with no decisive results scored by either side. One week later, on June 13, the squadron scored its first victory and had its first combat loss.

Additional losses followed on July 2, when a patrol from the 27th Aero entered what was considered its biggest fight to date. North-east of Chateau-Thierry, at 7:15 a.m., this group engaged nine Fokker D.VII's from *Jagdstaffels* 4 and 10 of *Jagdgeschwader* 1. A half-hour long battle ensued and four German planes were claimed by the Americans (two confirmed), while two pilots from the 27th Aero were listed as missing.[6] In fact, one had been shot down by *Leutnant* Ernst Udet to become his 39th victory.[7]

Frank Luke in leather flying coat, sitting on Nieuport 28. (Photo courtesy NASM.)

The following week, on July 9, the 27th Aero Squadron moved to a small aerodrome at Saints, a village about 8 km from Coulommiers. Soon after, a large number of replacement pilots arrived to compensate for the losses suffered to date. Two young men of opposite personalities were among the new-comers who appeared on July 25 – Second Lieutenant Frank Luke and First Lieutenant Joseph Wehner.

Little time was wasted on transitioning the replacement pilots into operational service, as illustrated by Luke's involvement with a patrol the next day, July 26. Despite having this mission cut short due to clouds and intermittent showers, it provided newer pilots like Luke with an opportunity to become acquainted with the sector. The weather cleared on July 28 and regular patrols resumed. Luke participated in six additional missions by July 31, but each was uneventful.

This pattern changed dramatically on the morning of August 1, when a patrol consisting of 18 planes departed on a protection mission for two Salmsons. This assignment, routine by nature, seemed destined for trouble. Only minutes into the flight, many of the squadron's Nieuports began to experience serious engine trouble. The pilots of these machines were compelled to drop-out of formation, one-by-one, until only six aircraft remained to meet and escort the Salmsons. This reduced escort was then engaged by a superior force of German aircraft. The results of this battle were horrendous – of the six Nieuport pilots who guarded the Salmsons, five were either killed or listed as missing.[8]

It is not clear which event ignited the trouble between Luke and other members of the unit. Likely it was a series of events. Nevertheless, one might consider the battle of August 1, in which the 27th Aero Squadron lost several of its most experienced pilots, save Lt. Sands, who was a recent replacement with limited experience over the front. Under normal conditions, Luke's brash exaggerations would have been brushed aside, perhaps jokingly. However, the losses incurred on that day created less tolerance towards such behavior. At that point it is likely Luke's comments, directed towards emphasizing his abilities and avenging squadron losses, were judged harshly as boastful behavior from an inexperienced youth. It is interesting to note that the following week, while participating on patrols of August 7, 9, 10, and 11, Luke consistently separated from the group and returned to the aerodrome alone, and well after everyone else.

Coinciding with Luke's activities in early August, 1918, the squadron's available aircraft increased considerably, as did the number of trial flights performed by the unit's pilots; both were good indicators that SPAD 13 aircraft had finally arrived in sufficient numbers. Unfortunately, the initial group of SPADs had a serious problem associated with a poorly housed reduction gear system. Apparently, with the slightest nick in the aircraft's propeller, a vibration would be generated of such magnitude that it eventually loosened enough critical pieces and caused the engine to miss or quit completely.

The vibration problem helped to set the stage for August 16, perhaps the most controversial day in Luke's

Major Harold E. Hartney. Prior to commanding the 27th Aero Squadron, Hartney served one year in the Canadian Infantry and nearly two years in the Royal Flying Corps, where he was officially credited with five victories. (Photo courtesy Nick Mladenoff.)

Nieuport 28s of the 27th Aero Squadron's 'A' Flight at Saints aerodrome. This line-up was photographed on August 1, 1918, about 25 minutes before six members of the flight were lost. Nieuport N6301 in the foreground with a light and dark colored number '16' on the upper wing, a concentric 'U'-shaped pattern on both sides of the cowling, and prominent commander's stripes on the fuselage, was probably flown by the 27th's commander, Major Harold E. Hartney. (Photo courtesy James Streckfuss.)

military career. On that morning, Hartney attempted to lead a group of SPADs on a protective patrol for a Salmson 2A2, only to have all other participating SPADs drop out due to engine trouble.[9] Back at the aerodrome, Luke was having similar problems with his SPAD. Nevertheless, he managed a take-off well after the patrol's departure and expected to locate the entire group between Fere-en-Tardenois and Fismes.

From this point, events become confused. According to witnesses at the aerodrome, Hartney returned from his flight, landed, and immediately summarized the results of his patrol to a listening crowd. Ironically, just as Luke's aircraft appeared over the aerodrome, Hartney commented that he had not seen Luke, or any hostile aircraft, during his flight. Considering Hartney was a highly experienced squadron leader – who had served nearly two years with the Royal Flying Corps and had five official victories – his words were generally taken as truth. Luke landed moments later and immediately summarized his own flight. In direct conflict with Hartney's comments, Luke indicated finding Hartney during his patrol. Then, to the astonishment of all, Luke explained that Hartney – despite his years of

Members of the 27th Aero Squadron gather behind Hoover's Nieuport No.8, serial N6157. This plane displays the squadron's 'eagle' insignia in its earliest form. (Photo courtesy Nick Mladenoff.)

Below right: With 'Jerry the mascot' looking on, Lt. Jerry Vasconcells (standing left) and Lt. Leo Dawson show the 27th Aero Squadron's 'eagle' insignia in its final form on a SPAD.

experience – had been closely followed by a group of five German aircraft. Apparently, as the dialog continued, Luke detailed how he flew towards one particular German aircraft which lagged behind the others. Then at a close range, he fired at this target twice – hitting it. After the second burst of gun fire, the enemy aircraft rolled onto its back and began to fall. Luke observed this aircraft continue down to a low altitude. "My last look at the plane shot down convinced me that he struck the ground for he was still on his back about 1500 meters below" is the wording of the combat report, yet he admitted not seeing the German aircraft crash. Luke, it then appears, was able to escape from the remaining enemy aircraft and made his way back to the aerodrome.

Luke's verbal description of the flight and subsequent combat report possessed one major anomaly – seeing Hartney is not mentioned in the latter. This lack of detail seems strange, especially if one considers a comment made by Hartney in later years: "I took Frank by the arm and walked him away from the others. From what he told me, the way he described it, and from the fact that I was there, I believed then, and always will believe, that he did shoot a German plane off my tail."[10]

Hartney's statement conveyed strong conviction, yet, it appears at the time that he was unable to sway others into believing it. If Hartney did not embellish the story, we are posed with an astonishing situation – judgment of Luke rested on two contradictory statements made by Hartney – one supporting Luke's claim, the other contradicting it.

Perhaps animosity toward Luke was a factor in the equation because the squadron embraced the latter with full vigor. The general opinion was that Luke had lied.

Luke may have embellished his story, too. Nevertheless, it is possible he shot an aircraft down, or at least forced one into an evasive maneuver, giving the impression it had been successfully attacked. To judge from his memoirs, Hartney half-expected and half-wished the theory would be confirmed. He later commented: "I used every resource I could muster, both before and after the Armistice, in trying to get him a confirmation, but that victory is unconfirmed to this day."[11]

The fact this claim was unconfirmed only strengthened the suspicions of Luke's critics. His popularity dropped to zero. Hartney commented that "Luke was a lonesome and despised man from that day until he brought down his first balloon near Mariculles on the St. Mihiel front on September 12th. His only friend was Joseph Wehner, who because of his German ancestry, had been detained and questioned several times by Secret Service agents before reaching the front. The common feeling of rejection brought these two closer together."[12]

Luke's combat report of August 16 reads as follows: "My machine was not ready so left an hour after formation, expecting to pick them up on the lines, but could not find formation. Saw Hun formation and followed, getting above into the sun. The formation was strung out leaving one machine way in the rear, being way above the sun directly behind. Opened fire at about 100 feet, keeping both guns on him until within a few feet of him, then zoomed away. When I next saw him he was on his back, but looked as if he was going to come out of it so I dove again holding both guns on him. Instead of coming out of it, he sideslipped off the opposite side much like a falling leaf and went down on his back. My last dive carried me out of reach of the other machine that had turned about. They gave chase for about five minutes and then turned back for I was leading them. My last look at the plane shot down convinced me that he struck the ground for he was still on his back about 1500 meters below. On coming home about our lines saw four E.A. started to get into the sun and above but they saw me and dove towards me. I peaked for home. Three turned back and the other came on. I kept out of range by peaking slightly, and he followed nearly to Coincy where he saw one of the 95th boys and turned about. The 95th man could have brought down the E.A. if he had realized quick enough that it was an E.A. The machine was brought down North East of Soissons in the vicinity of Joui and Vailly. Do not know the exact location as this being my first combat did not notice closely but know that it was some distance within German territory, for archies followed me for about ten minutes on my way

Lt. Joseph Wehner (standing to left) and Major Harold E. Hartney. These two men were perhaps the only true friends Luke had during his time at the front. (Photo courtesy Nick Mladenoff.)

back. My motor was fixed at Coincy and filled with gas and oil. Also found out that our formation had been held up by the Salmson that it was to escort and had just started. So left the ground to find them. Flew at about 5,000 meters from Soissons past Fismes, but did not see the formation. Saw one Salmson but no enemy E.A. Returned home."

On August 21, Major Hartney was given command of the First Pursuit Group, while the newly promoted Captain Alfred A. Grant took his place as CO of the 27th Aero

Frank Luke is shown with his mechanics in the 27th Aero Squadron, about mid-September 1918. Standing (left to right) Lou Flanner, unknown, Luke, and unknown. The radiator shell on this SPAD has not been painted blue (note contrast between darker camouflage and lighter underside colors on shell), suggesting this is not Bleriot-build SPAD 13 C1 No.26 shown opposite. (Photo courtesy Nick Mladenoff.)

Wehner, Roberts, and Cogrove by Frank Luke's SPAD, No.26, at Rembercourt, about September 1, 1918. Joseph Wehner was Luke's wingman and was shot down and killed in an attack on two German balloons on September 18. Ivan Roberts then became Luke's wingman but was also shot down eight days later while protecting Luke from several Fokkers. At the time William Cosgrove was the Armament Officer for the 27th Squadron. (Photo courtesy Nick Mladenoff.)

Squadron. Grant, a graduate of West Point, operated strictly by the book and, therefore, was not happy with the squadron's resident blacksheep – Frank Luke. Grant seriously considered court-martialing Luke more than once, despite Hartney's continued advice to show patience.

During this time military operations in the Chateau-Thierry sector quieted down, and the activities of the 27th Aero Squadron had little strategic importance. Records show that pilots spent a majority of their time brushing up on target practice and conducting formation flights. On August 29, 30, and 31, Luke and Wehner flew their training flights together – an indication of the strong friendship growing between them.

In early September 1918 the 27th Aero was assigned to the St. Mihiel sector. Occupying the air field at Rembercourt, they prepared for the next Allied drive which started at 5:00 a.m. on the morning of September 12. Luke was in the air at 7:29 a.m. – the unit's first patrol of the day – but soon left formation to chase three enemy aircraft. Luke was not able to engage these machines, yet he located and attacked an enemy balloon at Marieville. After conducting his third pass at the target, it burst into flames and quickly fell to the ground. Then Luke flew to the nearest American balloon company, knowing its members would have seen his attack. Upon landing, he jumped out of his SPAD with the propeller slowly ticking over, and had Lieutenants Fox and Smith, the most senior officers present, carefully read and sign his prepared confirmation note.

Confirmation by Lieutenants Fox and Smith eliminated any doubt about Luke's victory claim of September 12 – an important issue with him, since disbelief and sharp criticism had been expressed over his claim of August 16. The level of criticism Luke had endured until the confirmed victory of September 12 cannot be accurately assessed. However, the fact he personally typed a

confirmation form in advance for witnesses to sign indicates it was high. Another indication is the fact that Luke mentioned these details in his combat report. In other words, Luke knew he had destroyed an enemy balloon above Marieville, but more importantly, he wanted everyone else to know it too.

The Germans certainly knew what Luke had done above Marieville, especially *Leutnant* Willy Klemm, who had come down caught in the rigging of his burning balloon with a bullet wound near the heart. He died a few days later.

Luke's September 12 combat report reads: "Saw 3 E.A. near Lavignuelle and gave chase following them directly east towards Pont-a Mousson where they disappeared towards Metz. Saw enemy balloon at Marieville, destroyed it after three passes at it. Each within a few yards of the balloon. The third pass was made when the balloon was very near the ground. Both guns stopped so pulled off to one side. Fixed left gun and turned about to make one final effort to burn it, but saw it had started the next instant it burst into great flames and dropped on the winch, destroying it. The observer Joseph M. Fox who saw the burning said he thought several were killed when it burst into flames so near the ground. There was a good field near our balloons, so landed for confirmation. Left field and started back when my motor began cutting out. Returned to same field, and there found out my motor could not be fixed, so returned by motorcycle. Attached you will find confirmation from Lt. Fox and Lt. Smith. Both saw burning."

On September 14, Luke joined the morning patrol, which departed the airfield at 9:30 a.m. A short time later he spotted an enemy balloon strategically placed near Boinville and left the formation to attack it. As Luke dove down onto the balloon, he was quickly followed by Lieutenants Dawson and Lennon, who had also noted the huge prize. Luke's initial gun fire did not ignite the

Frank Luke with Bleriot-built SPAD 13 C1 No.26 of the 27th Aero Squadron, serial number unknown. This is the only photo that shows Luke with a plane that has some recognizable markings. And although it was not unusual for personnel to be photographed next to some plane other then their own, it more often than not signified ownership. This suggests that Luke flew plane No.26 in the squadron. American pursuit squadrons were organized into three flights and the planes were normally numbered in ascending order according to flight. The flights were further identified by having the radiator shell painted red, white, or blue. The aircraft in this photo appears to have a blue radiator shell which would designate the 3rd flight and would be consistent with the number 26.

explosive gas within the balloon. However, it did unnerve the observer, who leaped out of his basket, making good his escape by parachute. What followed can best be described as a series of undaunted attacks on the balloon as enemy anti-aircraft artillery fire began to burst close to the three flyers. In fact, in defiance of the exploding shells, Luke made at least one pass on the enemy artillery battery, scattering its crews. Moments later, the balloon, literally perforated by hundreds of bullets, simply deflated and fell to earth.

Luke submitted a report which stated: "Left formation at Abaucourt and attacked an enemy balloon near Boinville. Dove at it six times at close range. Had two stoppages with left gun which carried incendiary bullets, and after fixing both, continued the attack. After about 75 rounds being left in right gun, I attacked an archie battery at base of balloon. Am sure that my fire took effect as the crew scattered. After my first attack on balloon, the observer jumped. The last I saw of the balloon, it was on the ground in a very flabby condition. Confirmation requested."

Upon landing, Luke learned the squadron had been ordered to bring down a balloon in the vicinity of Buzy. He requested the assignment and asked Wehner to fly guard while he went after the enemy target. With the plan approved, the patrol took off at 2:30 p.m. Although discrepancies exist between Luke's and Wehner's combat reports, the story is generally the same. As agreed in advance, Luke broke formation and dove on the enemy balloon above Buzy. He quickly destroyed it, but he was then attacked by eight enemy fighters. In the resulting melee Luke's guns jammed, while several enemy pilots were raking his SPAD with machine-gun fire. Meanwhile, Wehner saw the balloon go up in flames and managed to locate Luke, who by this time was in a precarious situation. Without hesitation, Wehner flew into the enemy formation and shot down two German aircraft, undoubtedly saving Luke's life. The first enemy machine

Luke relaxing near the aerodrome, circa September 1918. (Photo courtesy Nick Mladenoff.)

Alfred Grand by his Nieuport 28 at Toul in June 1918. The 27th Aero Squadron was issued Nieuports in June and moved with the 1st Pursuit Group to Saints in July for the Battle of Chateau-Thierry. The 27th received its SPADs in August (Photo courtesy Nick Mladenoff.)

Wehner hit – a Fokker D.VII – fell off into a spiral sideslip and then spun into the ground. The second, an Albatros, turned and started towards the ground in a steep dive, possibly forced to land.

Luke's combat report reads: "I and Lt. Wehner were to leave with formation, dropping out at Buzy to attack enemy balloon. By orders of the C.O. on arriving at Buzy left formation and brought down enemy balloon in flames. While fixing my gun so that I could attack another nearby balloon, eight enemy Fokkers dropped down on me. Dove and pulled away from them. They scored several good shots on my plane. I saw Lt. Wehner dive through enemy formation and attack two enemy planes on my tail, but as my guns were jammed did not turn, as I was not sure it was an allied plane until he joined me later. You will find attached confirmation of balloon."

On September 15, Luke and Wehner were again instructed to attack enemy balloons south of Boinville. Despite last minute problems which prevented Wehner's participation, Luke took off with the patrol at the designated time – 4:20 p.m. – and as agreed broke formation over the target area.[13] Diving down alone, Luke's gun fire quickly hit two balloons which burst into flames. One of the victims, *Leutnant* Wenzel of *Ballonzug*

(Bz) 18 of Ballon Battalion 33, parachuted to safety.

Luke's combat report reads "I left formation and attacked an enemy balloon near Boinville in accordance with instructions and destroyed it. I fired 125 rounds. I then attacked another balloon near Bois d'Hingry, and fired 50 rounds into it. Two confirmations requested."

That same evening, Luke departed the aerodrome at 6:50 p.m. to attack yet another balloon. In pursuit of this mission, he flew through heavy anti-aircraft fire. Nevertheless, he was successful and the balloon exploded in flames at 7:50 p.m. However, on the return flight, Luke became disorientated in the fading light and was forced to land in a wheat field at Agers at 9:30 p.m. The next day he returned to his aerodrome.

Luke's combat report stated: "Patrolled to observe enemy activity. Left a little after formation, expecting to find it on the lines. On arriving there I could not find formation, but saw artillery firing on both sides, also saw light at about 500 meters. At first I thought it was an observation machine, but on nearing it, I found that it was a Hun balloon, so I attacked and destroyed it. I was archied

Luke sights his target while preparing to pull the trigger on what appears to be a captured German Maxim machine gun, circa September 1918. Although it was unusual to find this type of machine gun at American aerodromes, it was common for pilots to practice shooting at the gun range.

with white fire and machine guns were very active. Returned very low. Saw thousands of small lights in woods north of Verdun. On account of darkness coming on I lost my way and landed in a French wheat field at Agers at about 21h 30. Balloon went down in flames at 19h 50."

Luke's technique of attacking enemy balloons at dusk was considered very successful by the American Air Service, to the point it prompted an aggressive campaign against the German targets, with Luke and Wehner as the central figures. This assessment was substantiated on the evening of the September 16 when the two men set out again. This time Hartney and Grant hosted Colonel William Mitchell and his staff at the aerodrome for the scheduled evening show.

Another witness to the exploit that evening was Lieutenant Edward V. Rickenbacker of the neighboring 94th Aero Squadron. He later wrote that when Luke walked to his aircraft he pointed out the two German observation balloons to the east of the field, adding "Keep your eyes on these two balloons. You will see that first one there go up in flames exactly at 7:15 and the other will do likewise at 7:19."[14]

Lieutenant Rickenbacker admitted that many doubted Luke's words, but added "everyone gathered together out in the open as the time grew near and kept their eyes glued to the distant specks in the sky."[15] Richenbacker heard Hartney suddenly exclaim "There goes the first one!"[16] The young Lieutenant then added: "a tremendous flame lit up the sky exactly on the dot. Not a word was spoken as we glanced at our watches and then on the skyline, almost on the second the group yelled simultaneously as a small blaze first lit and then turned into a gigantic burst of flames."[17]

That night Luke and Wehner had simultaneously attacked the first balloon at the pre-arranged time; the actual explosion above Reville was closer to 7:05 p.m. Yet this was of little consequence to the hapless observer, who had jumped by parachute only seconds before the gigantic explosion engulfed his body in flames. The explosion had its effect on the two Americans, too. The intensity of the glare and the heavy anti-aircraft fire directed towards them caused the two flyers to loose track of one another. Wehner went for the balloon above Romagne next, only to see it go up in flames under Luke's attack at 7:19 p.m. Wehner then found and flamed an unscheduled third balloon just above the tree tops near Mangiennes at 7:35 p.m. The pair finally landed in the dark, supported by a crude system of lights which outlined the field.

Luke's combat report reads: "Patrol to straffe balloons-everything very carefully arranged. Lt. Wehner and I left aerodrome passing over Verdun. We attacked balloon in vicinity of Reville at 19h 03. Both Lt. Wehner and I shot a burst into it. It burst into flames and fell on observer who had jumped a few seconds before. We started for another balloon in vicinity of Ramagne. I attacked and destroyed it. It burst into flames on the ground burning winch. The anti-aircraft guns were very active scoring several good hits on my plane. The last I saw of Lt. Wehner he was going in a south-easterly direction after the first balloon went down. I shot at supply trains on my way back. Two confirmations requested."

Behind the German lines, *Leutnants* Finster and Heicke of *Ballonzug* 112, Ballon Battalion 20, also filed combat reports. Theirs testified to the demise of their balloon, despite the accurate anti-aircraft fire from *Leutnant* Roerich's *Flakzug* unit. Ironically, the *Flakzug's* accuracy was also noted in Luke's report. Yet Hartney's comment better illustrated their effect on Luke's and Wehner's aircraft when he indicated that "literally those planes had been all but shot out from under them. Each had at least 50 bullet holes and was useless for further service without a complete overhaul."[18] He added that in the course of only five days, "this was the fifth ship Luke had rendered unfit for further work on account of bullet holes."[19]

Luke, perhaps blinded by glory, seemed unconcerned with the damage his aircraft sustained. For in the same five days, it appears Luke, the boastful youth from Arizona, had transformed himself and Wehner – the suspected German spy from Massachusetts – into the most talked about fighting team in the Army Air Service. If true, First Lieutenant Kenneth Porter of the 147th Aero Squadron revealed a paradox when he expressed most of the First Pursuit Group's attitude toward Luke, even at the height of his fame: "Everybody knew him...everybody was against him. He always violated the rules. We all thought he was

Members of the 27th Aero Squadron stand around a captured German Maxim machine gun while Luke grips the triggers. (Photo courtesy Nick Mladenoff.)

screwy. We tolerated him, but we didn't want to be like him. Who wants to be nuts?"[20]

Despite the criticism, no one could dispute this team's phenomenal success. Luke had accounted for eight enemy balloons, while Wehner was credited with three balloons and two enemy aircraft in the course of assisting and protecting him. Unfortunately, their success was short lived, and in two days came to an abrupt end.

Luke rested on September 17, but the two were off again the next evening and successfully destroyed two German balloons near Labeuville. They were then surprised by a number of German Fokker D.VIIs, the main formation attacking Wehner, while two concentrated on Luke. Luke turned towards his two assailants and fired as they came head on, until only a few yards separated the planes. At that point one Fokker, apparently hit by gun fire, turned on its nose and fell to earth, crashing seconds later. The second German aircraft received another burst of gun fire from Luke and also fell into a dive. At this point, Luke was unable to find Wehner and, therefore, turned towards the Allied lines.

Unknown to Luke, Wehner had been shot down in his SPAD 13, probably by *Leutnant* Georg von Hantelmann of *Jasta* 15.[21] Struck by at least three bullets, Wehner crash-landed in the vicinity of St. Hilaire, and he died soon after.

Meanwhile, as Luke neared the front lines, he observed the airborne explosions of anti-aircraft fire and flew towards it to investigate. His flight path placed him southeast of Verdun; there he came upon a German reconnaissance plane – an LVG – being chased by two French SPADs from SPA 68. Although Luke's aircraft was extremely low on fuel, he rushed in to join the chase and helped to shoot the LVG down – its German crew, *Leutnants* Ernst Hohne and Ernst Schulz of *Flieger Abteilung* 36, were both killed in the attack.

One of the SPADs from SPA 68 was actually piloted by an American, *Adjutant* Reginald Sinclaire, who seemed slightly perturbed with Luke when he commented on this event: "We spent two hours stalking that two-seater, which just kept turning away whenever we presented a threat, then would return to his work when we seemed to give up. After three attempts we finally cut him off. At the same time, an American SPAD came up under his tail, also shooting, and landed where the German fell."[22]

Nearly out of fuel, Luke had little choice when he elected to land his SPAD. Nevertheless, this allowed him to personally view the wreckage of the German machine. And admittedly, Luke seemed somewhat excited around the wreckage – it did afford him his first opportunity to see a victim up-close. However, afterwards his mood became quite somber and he spent the night near Verdun, grieving for his lost friend, Joseph Wehner.

Luke's grieving continued into the next day, and Rickenbacker and Hartney drove to the front to retrieve him. Although Wehner's fate was still uncertain at that time, Hartney's comment best signified Luke's sense of despair when he stated "His eyes seemed already resigned to what he himself had seen."[23]

According to Hartney, Luke asked only one question and directed it to him: "Wehner isn't back yet, is he Major?"[24] Strangely, Hartney never revealed his reply. It seems impossible to imagine Hartney not responding to Luke's question. Perhaps he was trying to maintain the privacy of the moment, for little else is detailed.

Luke's combat report for September 18 reads as follows: "Lt. Wehner and I left the airdrome at 16h 00 to spot enemy balloons. Over St. Mihiel we saw two German balloons near Labeuville. We maneuvered in the clouds and dropped down, burning both. We were then attacked by a number of E.A. The main formation attacking Lt. Wehner who was above and on one side. I started climbing to join the fight, when two E.A. attacked me from the rear.

Below: Unidentified French soldiers examining the remains of a German L.V.G. shot down by Luke on September 18, 1918.

I turned on them, opening both guns on the leader. We came head on until within a few yards of each other, when my opponent turned on the second, shot a short burst and turned and went into a dive. I saw a number of E.A. above, but could not find Lt. Wehner, so turned and made for our line. The above fight occurred in the vicinity of St. Hilaire. On reaching our balloon line, flew East. Saw archie on our side, flew towards it and found an enemy observation machine. I gave chase with some other SPADs, and we got him off from his lines, and after a short encounter he crashed, within our lines, southeast of Verdun. Lt. Wehner is entitled to share in the victories over both the balloons. Confirmations requested, two balloons and three planes."

The five victories of September 18 raised Luke's total to 13; he was now the highest scoring American ace. He had also helped to make the 27th Aero Squadron the top-scoring outfit in the AEF.[25] In view of this outstanding achievement, Luke was recommend for a second Distinguished Service Cross (Oak Leaf Cluster), and on the night of September 20, the entire First Pursuit Group threw a party in his honor.[26] Although the party was intended to raise Luke's spirits, unfortunately, it achieved little. The next day, almost by force, he was sent on an extended leave to Orly and Paris, a gift from Hartney. This too, failed to improve Luke's depression and he returned before the entire leave had expired, anxious to resume combat.

The next day, September 26, Luke and Second Lieutenant Ivan A. Roberts teamed up for a patrol at 5:15 p.m. They set out for the enemy balloon line, but never made it, being jumped by at least five Fokkers in the vicinity of Consenvoye and Sivry. Although Luke succeeded in shooting one enemy aircraft down, two more Fokkers got on his tail and caused him to lose sight of Roberts. At this point, Luke was plagued by jamming machine guns, yet he fought off the enemy aircraft. Roberts was not as fortunate; he was shot down and made a POW, most likely the result of gun fire from *Leutnant* Franz Büchner of *Jasta* 13.[27]

Luke's combat report for September 26 reads: "On patrol to straffe balloons in vicinity of Consenvoye and Sivry, I attacked with two others a formation of five (5) Fokkers. After firing several short bursts, observed the Hun go down out of control. While at 100 meters I was attacked by two (2) E.A., so I did not see the first E.A. crash. I turned on the other two who were on by tail, getting on the tail of one, but guns jammed several times, and after fixing both could only shoot short burst on account of the several stoppages. One (1) confirmation requested. The last I saw of Lt. Roberts, who was on this patrol with me, was in combat with several Fokkers in the vicinity of Consenvoye and Sivry."

When Luke learned that Roberts was missing under the

Frank Luke standing beside the remains of a German L.V.G. from *Flieger Abteilung* 36. Luke shot this aircraft down on September 18, 1918, for his 13th victory.

same circumstances as Wehner, namely guarding him, Luke left the aerodrome without permission. He returned on September 27, and was severely reprimanded by Captain Grant – apparently with little effect. The next morning, while Grant and Hartney were discussing Luke's future, he took-off without permission. Moments later, Luke flew across the lines at only 500 meters. He then successfully destroyed an enemy balloon at Bantheville. His mission complete, Luke landed at a French aerodrome and spent the night. He returned to his own airfield the next morning and filed a very brief combat report which stated: "I flew north to Verdun, crossed the lines at about five hundred meters and found a balloon in its nest in the region of Bantheville. I dove on it firing both guns. After I pulled away it burst into flames. As I could not find any others I returned to the airdrome. One confirmation requested."

Hearing of Luke's return, Grant immediately grounded him. Apparently Grant felt the need to discipline Luke for his recent actions despite the victory of September 28. It also seems apparent Luke thought little of Grant, or at least his orders. Only moments after their discussion ended, he climbed into a SPAD 13 and departed the aerodrome, landing at the advanced field near Verdun a short time later with engine trouble.[28]

At this point, Grant appeared desperate over his inability to control Luke's behavior and attempted to have Luke placed under arrest. Surprisingly, this action was prevented by Hartney, who later admitted lying when he told Grant he had already given Luke permission to fly out of the Verdun field for a balloon strafe.[29] Apparently, Hartney had tried once again to defuse the mounting tension caused by Luke's undisciplined behavior.

Hartney did, however, call the advanced field, then under the control of First Lieutenant Vasconcells, and issued absolute orders that prevented Luke's plane from taking off until after 5:00 p.m. In the meantime, Hartney flew to the Verdun field under the pretense of escorting a new pilot, yet he would later write he wanted to view the situation himself. Hartney commented that Luke was pacing around the field like a caged lion, "almost wild with impatience."[30] This was confirmed by Luke's attempt to depart half an hour before he was scheduled to fly. However, before he succeeded, Hartney told Vasconcells to pull Luke out of his SPAD and to warn him about obeying his orders. Hartney later wrote that "in a moment Luke's propeller stopped and Vasconcells made him get out of his ship. Luke looked at me sheepishly,

Close-up of Luke after he shot down German L.V.G. on September 18, 1918. Luke was still uncertain of Wehner's fate. Yet, Hartney's comment best illustrated Luke's facial expression at this point: "His eyes seemed already resigned to what he himself had seen."

grinning. I shook my fist at him. Frank knew he couldn't get away with it."[31]

Ironically, that was the last gesture Luke saw from Hartney, perhaps one of the only true friends he had during his time at the front. Hartney had, against all rules of military discipline, continuously supported Luke's unorthodox methods and afforded him the opportunity to transform himself into the phenomenal success he had boasted of becoming.

After Hartney departed the field, Luke attempted to take-off at 5:00 p.m., but landed again with more engine trouble. He succeeded during a second try near 7:00 p.m.[32] Once in the air, Luke flew low over the 7th American Balloon Company at Souilly and dropped a message in a cylinder: "Watch for burning balloons. Luke."

Frank Luke never returned from his mission on September 29. He was listed as missing. However, the 7th Balloon company at Souilly, did observe the destruction of three enemy balloons near Avocourt. The noted times were 5:05, 5:06, and 5:12p.m.

On October 19, General Order No. 17 was issued by Headquarters, First Army, Air Service, AEF, and officially credited Luke with the destruction of the three balloons on September 29.

Two weeks later, on November 3, Captain Alfred A. Grant submitted a recommendation to award Luke the Medal of Honor "for gallantry in action and for exceptionally meritorious service." In support of the recommendation, the document listed each of Luke's victories. Ironically, Luke's unconfirmed victory claim of August 16 was the first entry. However, if one considers the level of criticism Luke endured because of this claim, it would not be surprising to see it listed if members of the squadron suspected Luke was dead and felt remorse.

The actual facts surrounding Luke's death did not surface until January 3, 1919. On this date, Captain Chester E. Staten, a Graves Registration Officer, addressed a letter to the Chief of Air Service, referencing the grave of an unknown American aviator. It reads in part: "Units of this service have located the grave of an unknown aviator killed on Sunday, September 29, 1918, in the village of Murvaux.

"From the inspection of the grave and interview held with inhabitants of this town, the following information was learned in regard to the heroism of this aviator. Any assistance you can furnish us that will enable us to properly identify this body will be greatly appreciated. The following might assist you in gaining for us this information. Reported as having light hair, young, of medium height and of heavy stature. Reported by the inhabitants that previous to being killed this man had brought down three German balloons, two German planes, and dropped hand bombs killing eleven German soldiers and wounding a number of others. He was wounded himself in the shoulder and evidently had to make a forced landing. Upon landing opened fire with his automatic and fought until he was killed. It is also reported that the Germans took his shoes, leggings, and money, leaving his grave unmarked."

Three days later, on January 7, an American Red Cross inter-office letter provided addition information concerning the still unidentified aviator:

Luke (standing to left) with unidentified aviators, on leave in September 1918. (Photo courtesy Nick Mladenoff.)

"This officer was killed at Murvaux (5 kilometers east of Dunsur-Meuse) on Sunday, September 29, 1918. The Germans stripped him of all identification, but Captain McCormick of the 301st Graves Registration station at Fontains near Murvaux was so interested in the story told by the French people of Murvaux concerning the death of this aviator that he exhumed the body and stated that it was that of a man of medium height, heavy set and with light hair. On his wrist he found an Elgin watch #20225566, which was under the sleeve of his combination and which the Germans who had stripped him of all papers and identification marks had evidently missed."

"The description of this aviator by Captain McCormick and the fact that Lieut. Frank Luke dropped a note to a balloon company that day stating he was going to shoot down the balloons which were shot down make it almost certain that this officer was 2d Lieut. Frank Luke, Air Service, whose nearest relative is Frank Luke, 2200 West Monroe St., Phoenix, Arizona. If the Air Service wishes to check this case, it is suggested that a representative of the Air Service be sent to Murvaux and obtain sworn statements from French people of that village."

On January 15, the following affidavit was obtained from the people of the town of Murvaux:

"The undersigned, living in the town of Murvaux, Department of the Meuse, certify to have seen on the 29th September, 1918, toward evening an American aviator, followed by an escadrille of Germans in the direction of Liny, near Dun (Meuse), descended suddenly and vertically toward the earth, then straighten out close to the ground and fly in direction of Briere Farm, near Doulcon, where he found a German captive balloon which he burned. Then he flew toward Milly (Meuse) where he found another balloon, which he also burned, in spite of incessant fire directed toward his machine. There he was

apparently wounded by a shot from rapid fire cannon. From there he came back over Murvaux, and still with his machine guns he killed six German soldiers and wounded as many more.

"Following this he landed, and got out of his machine, undoubtedly to quench his thirst at a nearby stream. He had gone some 50 yards, when seeing the Germans come toward him, still had the strength to draw his revolver to defend himself, and a moment after he fell dead, following a serious wound he received in the chest.

"Certify equally to have seen the German commandant of the village refuse to have straw placed in the cart carrying the dead aviator to the village cemetery. This same officer drove away some women bringing a sheet to serve as a shroud for the hero, and said, kicking the body – "Get that out of my way as quick as possible."

"The next day the Germans took away the aeroplane, and the inhabitants also saw another American aviator fly very low over the town apparently looking for the disappeared aviator."

"Signatures of the following inhabitants: Perton, Valentine Garre, René Colon, Gustave Carre, Auguste Cuny, Geon Henry, Henry Gustave, Cortine Delbart, Eugene Coline, Gabriel Didier, Odile Patouche, Camille Phillips, Richard Victor, Voliner Nicholas.

"The undersigned themselves placed the body of the aviator on the wagon and conducted it to the cemetery: Cortine Delbart, Voliner Nicholas.

"Seen for legalization of signatures placed above, Murvaux, Jan. 15, 1919." This affidavit was signed by the Mayor of Murvaux – Auguste Garre – and was stamped with the seal of Murvaux.

The American Air Service considered the investigation complete. Subsequently, Luke's body was again buried at Mervaux. There an American officer from the air service made arrangements to have a simple cross erected over

Frank Luke, Jr. On his left wrist, just visible under his sleeve, is the Elgin watch (#20225566) which was one of the major identification features when Luke's body was exhumed.

Luke's grave. It carried the following inscription: 2nd Lt. Frank Luke, Pilot 27th Aero Squadron, 19 victories, killed in action September 29, 1918.

The 19 victories listed on Luke's cross was an error; his victory of August 16 was never confirmed. Therefore, Luke's official score was 18, despite the fact French civilians reported to have seen Luke down three balloons and two aircraft. Apparently, only the enemy balloons were witnessed by the American balloon company and officially credited to his score.

Luke's first German balloon on September 29, 1918, had been Ballon Nr. 35, attached to the German *5 Armee* at Cote-Saint-Germain; the second belonged to *Ballonzug* 64 at Briere Farm. The third, at Milly, belonged to a Ballonzug commanded by *Leutnant* B. Mangels, who subsequently got into a dispute with *Leutnant* G. Roesch, commander of the nearby *Flakzug*, as to who brought down the American balloon buster.[33] Roesch ended up conceding the honors to the machine-gunners of Mangels' balloon company. Even so, it ultimately took a shootout at close range to finish Frank Luke.

On January 25, 1919, Lieutenant Colonel Aleshire, commanding officer of the First Air Depot filed a recommendation for Luke to receive the Medal of Honor posthumously. It was awarded that year.

Luke's other awards included the Distinguished Service Cross with Oak Leaf Cluster, the Italian *Croce di Guerra,* the Aero Club Medal for Bravery, and as the first graduate of Rockwell Field to score an aerial victory, Rockwell's Margarita Fisher Gold Medal.

Francis Edward Ormsbee, Jr.

Francis Edward Ormsbee, Jr., circa 1919. A Medal of Honor ribbon is visible on the right breast of his uniform. The emblem on his right sleeve indicates his rank to be Chief Machinist's Mate, Aviation Section. (Photo courtesy National Archives.)

Such awards recognized bravery in saving life, and acts of valor performed in submarine rescues, boiler explosions, turret fires, and other types of disaster unique to the naval profession. But in each case, the individual distinguished himself with conspicuous gallantry or intrepidity, at risk of life above and beyond the call of duty.

In fact, 40 years after the Medal's conception, the Navy reinforced its views towards "bravery in the line of the naval profession," by specifying in the Congressional Act of March 3, 1901, "that any enlisted man of the Navy or Marine Corps who shall have distinguished himself in battle or display extraordinary heroism in the line of his profession shall, upon recommendation of his commanding officer, approved by the flag officer and the Secretary of the Navy, receive a gratuity and the medal of honor..."

The next time the Navy addressed the issue was in a legislative act submitted to Congress on July 9, 1918. Again, the Navy seemed concerned with acknowledging "bravery in the line of the naval profession," but also expressed the desire to elevate the medal's status as a combat award.

As a result, the Navy created a second Medal of Honor to distinguish the two categories: the original Medal design (five-pointed star) was maintained as an award for non-combat service; and the 'Tiffany Cross' design was established as an award for "any person who, while in the Naval service of the United Sates, shall, in action involving actual conflict with the enemy, distinguish himself conspicuously by gallantry and intrepidity at the risk of his life above and beyond the call of duty and without detriment to the mission of his command or the command to which attached."

In addition to providing new regulation for the Medal of Honor, the Act of July 9, 1918, also established the distinguished-service medals and navy crosses. This thereby became the genesis of what has been called the "Pyramid of Honor," a hierarchy of 12 military decorations awarded for combat valor and meritorious service, at the top of which is placed the Medal of Honor.

Yet in Ormsbee's case, the Act of July 9, 1918, is of little consequence. It was not approved by Congress until February 4, 1919, while Ormbsee's Medal of Honor recommendation was approved by the Secretary of the Navy two months earlier, on December 3, 1918. And in either case, Ormsbee's award would have been the original Medal design (five-pointed star), because the newer congressional act maintained it as an award for non-combat service.

Unfortunately, Ormsbee's service records have been lost. Yet bits and pieces tell us he was born in Providence, Rhode Island, on April 30, 1892. He was the first of two children born to Francis and Sarah Ormsbee. His mother was an Irish immigrant, while his father, a native of Providence, worked as a carpenter.

When the Secretary of the Navy approved General Order 436 on December 3, 1918, Francis Edward Ormsbee, Jr. became the first aviator awarded the Medal of Honor.[1] This unique distinction is made even more interesting by the fact Ormsbee's Medal was awarded for an act of bravery not involving conflict with an enemy force, and is therefore referred to as a "peacetime" Medal of Honor.

As controversial as that may seem, the reason for bestowing a "peacetime" Medal of Honor on an individual was established with the Congressional Act that created the award on December 21, 1861, having specified gallantry in action "and other seaman-like qualities" as the basis for the Medal. For this reason the Navy could, and did, award the Medal of Honor for bravery in the line of the naval profession. In fact, over the years the Navy has issued 180 "peacetime" Medals of Honor.

F.E. ORMSBE.
A.C.M.M.

Francis Edward Ormsbee, Jr.,
late February 1923. The
silver wings pinned on his
right breast show he had
qualified as a "Chief Aviation
Pilot" (CAP). (Photo courtesy
National Archives.)

Francis Ormsbee, Jr., had attended school in Providence, but dropped out to enlisted in the Navy at an early age. By the time the United States entered World War One, he was listed as a Chief Machinist's Mate and attached to the Naval Air Station (NAS) at Pensacola, Florida.

As a chief petty officer, Ormsbee held a position of substantial importance. In fact, Chief Machinist's Mates managed the daily maintenance and use of every seaplane at the station, to the level that they managed every needed function from the changing of oil to major construction. During the war, Ormsbee would have been extremely active in these day-to-day functions, because the Navy had concentrated all its heavier-than-air pilot training at Pensacola.

Naturally, flying accidents had become routine at the base. However, the crash of a Burgess N-9 Seaplane in Pensacola Bay on September 25, 1918, is unique due to the fact Ormsbee displayed extraordinary heroism in the line of his profession, and was awarded the Medal of Honor as a result.

The following is the "Record of Proceedings of a Board Of Investigation" which convened at NAS Pensacola, Florida, days later. It is quoted in its entirety:

NAS, Pensacola, Florida with Curtiss trainers lined up, January 27, 1917. The blue anchor insignia, visible on the tail of each aircraft, was first adopted by the Navy 1916. These hangars are still in use today. (Photo courtesy Noel Shirley.)

****

Record of Proceedings Of A
Board Of Investigation
Convened at
U.S. Naval Air Station, Pensacola, Florida.
By Order of
THE COMMANDANT, U.S. NAVAL AIR STATION
PENSACOLA, FLORIDA.

FROM: Commandant
TO: Lieut. Nugent Fallon, U.S.N.R.F.

SUBJECT: Board of Investigation of Wreck of Seaplane No. 2422, that occurred in Pensacola Bay, Pensacola, Florida, on September 25, 1918, at the U.S. Naval Air Station, Pensacola Florida.

1. A Board of Investigation consisting of yourself as Senior Member and Ensign Augustus B. Richardson, U.S.N.R.F., and Ensign Carl E.Finch, U.S.N.R.F., as additional members will convene at 10:00 A.M., Friday, September 27th, 1918, for the purpose of inquiring into and reporting upon the wreck of Seaplane No.2422 that occurred in Pensacola Bay,

Pensacola, Florida, September 25, 1918.

2. The Board will make a thorough investigation of all the circumstances attendant to the above mentioned wreck and upon the conclusion of its investigation will report the facts established thereby, the amount of damage to the Seaplane, and the Board's conclusion as to the responsibility for the wreck, and will forward in addition to the original record of proceedings, a partial copy covering material in accordance with Section 593, Naval Courts and Boards.

3. The attention of the Board is particularly invited to Section 589 and 511, Naval Courts and Boards.

F.M. Bennett.

U.S. Naval Air Station,
Pensacola, Florida,
Friday, 27 September 1918.

The Board met at 10:00 A.M.

In addition to training planes, Pensacola was equipped with several large Curtiss flying boats for active patrols off the coast of Florida. This photo shows three large Curtiss H-12s; serial numbers 4061 and 3536 are visible. (Photo courtesy Noel Shirley.)

Present:
Lieutenant N. Fallon, U.S.N.R.F., Senior Member, and
Ensign A.B. Richardson, U.S.N.R.F., Member, and
Ensign C.E. Finch, U.S.N.R.F., Recorder and Member.

Grady Deen, C.M.M. (A) U.S.N., was called as a witness. Examined by the Recorder:
1. Q: "Tell what you know regarding the wreck of seaplane #2422, at Pensacola, Florida, on the morning of 25 September 1918."
A: "Plane 2422 was reported as having been in bad weather over at Camp Saufley and needed being examined. Plane was inspected by myself and found to be a little out of alignment. This was on Tuesday, September 24th, and I relined the plane. The plane was finished and reported ready for flight."

2. Q: "Did you examine all controls?"
A: "Yes, sir."

3. Q: "Then the last inspection of the plane was made on Tuesday?"
A: "It was examined by the Tuesday "Night Crew" early Wednesday morning."

4. Q: "What is your duty?"
A: "Structural Officer of Squadron I."

5. Q: "Did the night crew report the plane as in perfect condition?"
A: "Ready for flight."

The Board did not desire further to examine this witness. He verified his declarations and withdrew.

Ensign Wallace C. Green, U.S.N.R.F., Commander Division A, Squadron I, was called as a witness. Examined by the Recorder:
1. Q: "Tell what you know regarding the wreck of seaplane #2422, at Pensacola, Florida, on the morning of 25 September 1918?"
A: "I was in charge of the flying over at the Island the day of the accident. At 9:45 A.M. plane #2422, Ensign Thomas C. McCarthy, pilot, P.B. Parsons as student gunner, left the beach at Camp Saufley on a routine gunnery hop. At approximately 10:10 A.M., men at the Island shouted that the plane had fallen near sector III range boat. Shortly afterwards I went to the scene of the accident, landed near the wrecked seaplane and noticed that the S.P. boat had a rope attached to the plane and was trying to lift the wreckage out of the water by using its anchor davit. I stayed on the scene of the accident until I saw the pilot's body taken from the plane and handed to the men in the sea-sled. I then returned to the Island to write out a report of the accident and carry on the usual work. The student

N-9 floatplane, serial 2466 taking off. This type was the standard floatplane trainer of the U.S. Navy during World War One. The emblem appearing aft of the serial numbers "2466" is most likely a squadron designation emblem. In May 1918, the Chief of Naval Operations granted permission for NAS Pensacola, Florida, to place squadron designating marks on seaplanes used for training purposes only. However, these marks had to be removed if the aircraft was transferred. (Photo courtesy Noel Shirley.)

gunner had been removed from the wreckage before I reached the scene of the accident."

2. Q: "Who had the plane previous to Ensign McCarthy?"
A: "The same pilot."

3. Q: "Did he test the controls previous to the flight and did he make any complaint of the condition of the plane between flight?"
A: "Ensign McCarthy stated that the plane was in good flying condition. He did not try the controls to my knowledge, but had used the controls in flying to the Island from the Station."

4. Q: "How much flying experience had Ensign McCarthy had?"
A: "35 hours and 35 minutes solo work."

5. Q: "How long had he been a gunnery pilot?"
A: "From September 19th, 1918 to time of accident, during which time he put in 7 hours and 10 minutes."

6. Q: "What were the weather conditions at the time of the accident or previous to the time of the accident?"
A: "At the time of the accident the air was slightly bumpy at 1000 feet at which height McCarthy was flying."

The Board did not desire further to examine this witness. He verified his declarations and withdrew.

Lt. (j.g.) C.S. Dunner, U.S.N.R.F., Medical Corps, was called a witness.
Examined by the Recorder:
1. Q: "Tell what you know regarding the wreck of seaplane #2422, at Pensacola, Florida, on the morning of 25 September 1918."
A: "While at the wet basin a sea-sled came in with P.B. Parsons, Elec.3c.Radio, whom I had transferred to stretcher in the following condition: He was in a moderate amount of shock, suffering severe pain. He had a posterior

dislocation of the left leg at the hip. Compound fracture of the nose, deep laceration over the left malar bone and one over the upper lip. In my opinion there were no internal injuries."

The Board did not desire further to examine this witness. He verified his declarations and withdrew.

Ensign James H. Hawkins, U.S.N.R.F., Flight Officer, was called a witness.
Examined by the Recorder:
1. Q: "Tell what you know regarding the wreck of seaplane #2422, at Pensacola, Florida, on the morning of the 25th. of Sept. 1918."
A: "I was Flight Officer on September 25th., 1918. I did not see the crash, but did see the disturbance in the water, perhaps a few seconds after the crash. The Dispensary and Aide for Operations were notified immediately. The Boat Officer was ordered to send the duty boat to the crash, which was in Sector III. The range boat evidently saw the crash and was alongside one minute after it happened. A barge was in the vicinity of the crash and it put about and was along-side in three minutes. At the time of the crash the sub-chaser was leaving the Station and it proceeded to the crash and stood by. The range boat returned with the student gunner and he was given over to the care of the Doctor, who was standing by at the wet basin. The duty boat instead of going in Sector III went out in Sector II, and failed to find the wreck. It returned to the wet basin, found out where the wreck was and proceeded to the scene. Another doctor was sent to the wreck in a speed boat."

2. Q: "What was the condition of the weather this day?"
A: "The condition of the air was evidently satisfactory, but the water was very rough and flying was later discontinued on account of this rough water."

3. Q: "Where was the wrecking barge at the time of the accident?"
A: "The wrecking barge was over near Santa Rosa Island,

N-9, serial 2561, at NAS Pensacola, Florida, September 1918. This training aircraft carries a "winged spade" emblem on its fuselage.

opposite the Station, working on the wreck of flying boat #779."

The Board did not desire further to examine this witness. He verified his declarations and withdrew.

Francis E. Ormsbee, C.M.M. (A) U.S.N., was called as a witness.
Examined by the Recorder:
1. Q: "Tell what you know regarding the wreck of seaplane #2422, at Pensacola, Florida, on the morning of 25 September 1918."
A: "I was out on a material test with Ensign J.A. Jova, U.S.N.R.F., in machine #2481, headed for Camp Saufley, when I saw a machine at an altitude of about 500 feet in a spinning nose dive. I was going to call Mr. Jova's attention to it when it came out of the spin into a straight dive and I assumed that the pilot would pull it up in time but he didn't and it crashed right into the bay. I called my pilots attention to it and we started down for the wreck. On landing near the wreck I dove over and swam to it, arriving before the speed boat, but found the pilot and gunner both submerged. The fuselage was turned nearly or partially downward, but not enough to prevent me from getting the gunner's head above water and I held his head above water until the speed boat arrived. I called for two men who jumped over immediately and they held him while I unfastened his safety belt and they took him on board the speed boat. In the meantime a barge had come and the speed boat shoved off with the injured gunner. I then dove under to locate the pilot and found that he was jammed in so bad by the wreckage that to extract him it would be necessary to at least partially lift the fuselage. I called for a rope and by going under the water fastened it around the fuselage, back of the pilots seat, and got all hands on board the barge to try and lift it up. We could not succeed and tried backing the barge up in hopes of lifting the fuselage to the surface in that way. It could not be done, although we continued trying until the S.P. boat

run its nose right up to the wreckage and cast us a block and fall off the anchor davit which we went under and fastened to the rope already tied around the fuselage. All hands hoisted the fuselage to the surface, which brought the pilot's head above water. This was about 20 minutes after the time of the crash. We could not extract the body, however, until we chopped the strut away with an ax, which we got from the S.P. Boat and had loosened his feet from the wires and wreckage and fastened a rope around his body and the S.P. Boat handed it over to the speed boat which had the Doctor in it. They took him on board and started working on him."

2. Q: "How long after the accident did the first boat arrive?"
A: "About a minute."

3. Q: "How did Ensign Jova reach the wreck?"
A: "Ensign Jova immediately nosed his machine down on a steep dive by skillful flying, landing almost beside the wreck and I had prepared to go over-board while coming down and dove in shortly after the plane landed."

4. Q: "Who were the men who dove to your assistance from the sled?"
A: "G.L. Blakemore, Printer First Class (A) U.S.N.R.F., and another man who got back into the boat almost at once on account of the cold."

5. Q: "What was the condition of the plane?"
A: "Completely demolished with the exception of the engine."

The Board did not desire further to examine this witness. He verified his declarations and withdrew.

Lieut. (j.g.) Toby Anthony Greco, U.S.N.R.F., Medical corps, was called as a witness.
Examined by the recorder:

A close-up of Curtiss H-12, serial 3536, NAS Pensacola, circa September 1918.

1. Q: "Tell what you know regarding the wreck of seaplane #2422, at Pensacola, Florida, on the morning of the 25th. of September 1918.

A: "I was called to attend the accident of Ensign Thomas C. McCarthy U.S.N.R.F., September 25th 1918. On reaching the scene of the accident the above named was still submerged, having been caught in the wreckage of the seaplane. After 20 minutes of submersion the body was taken aboard the speed boat and resuscitation immediately begun, strychnine sulfate given hypodermically. Patient brought to the dispensary and artificial respiration continued for two hours. External heat applied. At the end of that time there was still no sign of life. Pronounced dead at 12:15. Patient also showed a compound fracture of left leg. As to cause of death, drowning, due to fatal submersion.

The board did not desire further to examine this witness. He verified his declarations and withdrew.

There were no further declarations to be introduced nor anything further to be offered by the recorder, and the Senior Member announced that the investigation was closed.

The Board, after maturely deliberating upon the declarations recorded above, and from an examination of the wreck, finds as follows:

1. That seaplane #2422, Burgess N-9 Hispano-Suiza engine, crashed in Pensacola Bay, Pensacola Florida, September 25th, 1918, at about 10:10 A.M., from a spinning nose dive.

2. That at the time of the accident seaplane #2422 was engaged in a properly authorized and fully equipped flight for machine gun target practice.

3. That an inspection had been made of this plane by a qualified and experienced structural officer shortly previous to the flight and that it was found in a satisfactory and safe condition.

4. That the pilot in charge at the time of the accident was the only one that had flown the plane subsequent to this inspection and that he made no complaint as to its condition upon landing previous to the last flight.

5. That particularly prompt and efficient aide was rendered with the equipment available, subsequent to the accident.

6. That the plane went into a spin through causes unknown at about 500 feet. It was extricated from this spin, but at such a low altitude that it crashed into the bay, before completely recovering its equilibrium.

7. That seaplane #2422 was completely demolished with the exception of the engine, which was salvaged.

8. That Ensign Thomas C. McCarthy, USNRF., lost his life in this accident and that Parsons, P.B. Elec.3c., Radio, was very seriously injured.

9. That highly commendable and courageous assistance

Curtiss N-9 trainers at NAS Pensacola, Florida, 1918. (Photo courtesy Noel Shirley.)

was rendered by Ensign J.A. Jova and Francis E. Ormsbee, C.M.M. (A) U.S.N., and G.L. Blakemore, printer 1c., (A) USNRF., and that the work of the former two men undoubtedly saved the life of the student gunner.

10. That the duty boat containing the pullmotor and the emergency equipment proceeded to the wrong sector and had to return to its base for further instruction.

CONCLUSIONS: The accident was due to lack of flying ability on the part of the pilot.

RECOMMENDATIONS: That barges and S.P. boats assigned to Air Stations be equipped with small hoisting devices for lifting planes out of water enough to enable crews to be extracted before they are drowned.

****

At the time of the incident, the Congressional Act of March 3, 1901 was still in effect. It specifically stated the Medal of Honor could be awarded to any enlisted man of the Navy who "displayed extraordinary heroism in the line of his profession." In Ormsbee's case, a board of investigation had concluded his actions on September 25, 1918, were highly "commendable and courageous," adding that his assistance "undoubtedly saved the life of the student gunner." Based on these findings, a recommendation to award Ormsbee the Medal of Honor was put into motion in October 1918, and approved by the Secretary of the Navy on December 3, 1918.

Ormsbee's citation reads in full: "For extraordinary heroism while attached to the Naval Air Station, Pensacola, Fla., on 25 September 1918. While flying with Ensign J.A. Jova, Ormsbee saw a plane go into a tailspin and crash about three-quarters of a mile to the right. Having landed near by Ormsbee lost no time in going overboard and made for the wreck, which was all under water except the two wing tips. He succeeded in partially extricating the gunner so that his head was out of the water, and held him in this position until the speedboat arrived. Ormsbee then made a number of desperate attempts to rescue the pilot, diving into the midst of the tangled wreckage although cut about the hands, but was too late to save his life."

Ormsbee remained active in naval aviation after World War One. On February 20, 1923, he qualified as a "Chief Aviation Pilot" (CAP). Unfortunately, he was killed in a airplane crash at Ardmore, Oklahoma, on October 24, 1936. He was buried six days later at St. Francis Cemetery, Providence, Rhode Island. At the time of his death, he was married to Elvine Gristie, but it is unknown if they had any children.

Above: Burgess N-9 #2452. (Photo courtesy National Archives.)

Right: CAP Francis Ormsbee, circa 1928. (Photo courtesy National Archives.)

Below: Curtiss R-9 floatplane. On Feb. 13, 1917, Captain Francis T. Evans, USMC, performed what was perhaps the most daring feat by any naval aviator up to that time when he looped a Curtiss N-9 from 3,000 feet not just once, but twice. He then went into a spin and pulled out safely. For this contribution to aviation and the safety of flight, Capt. Evans received the Distinguished Flying Cross. (Photo courtesy National Archives.)

Edward Vernon Rickenbacker

First Lieutenant Edward V. Rickenbacker. With 26 victories he finished the war as "America's Ace of Aces." By date of action he was awarded the Medal of Honor (25 September 1918), the Distinguished Service Cross (29 April 1918) with nine Oak leaf clusters (May 17, May 22, May 28, May 30, May 14, Sept. 14, Sept. 15, Sept. 25, Sept. 26), the French *Croix de Guerre* (May 9, 1918), with three palms and *Chevalier de la Legion d'Honneur*. (Photo NASM.)

Edward Vernon Rickenbacker ended the First World War with 26 victories and the title "America's Ace of Aces."[1] An impressive accomplishment, considering many of these victories were obtained through repeated acts of extraordinary heroism, resulting in Rickenbacker being awarded the Distinguished Service Cross (DSC), with nine Oak Leaf Clusters. In fact, his DSC count remains the highest ever awarded to one individual. Perhaps more significant is that Rickenbacker achieved his victories in less than six months at the front. Yet ironically, it took 12 years for the United States to award him the Medal of Honor for his "conspicuous gallantry and intrepidity above and beyond the call of duty." It was finally presented to him by President Herbert Hoover at Bolling Field, Washington, DC, on November 6, 1930.

Rickenbacker was born October 8, 1890 in Columbus, Ohio, one of seven children born to Swiss immigrant parents (the family used the more Germanic spelling, Rickenbacher, prior to America's entry in the war). Due to his father's sudden death, he left school at the youthful age of 13 and began working in a glass company in an effort to support the family. Rickenbacker would later state: "I had never formally quit school; rather, I had simply stopped going."[2] He continued his education by studying mechanical engineering and drafting through a correspondence course, and by 1906 he was able to road-test automobiles. Motor racing followed and it brought Rickenbacker considerable fame and income; he became one of the country's premier race car drivers, earning nearly $60,000 in 1916 alone.

That same year, Rickenbacker traveled to England with plans of exporting several English racing automobiles to America, specifically those built by the Sunbeam Motor Works. However, the economics of war ended the plan before it took hold. To make matters worse, due to his Teutonic name he was suspected of being a German spy and temporarily held by Secret Service officials. Although eventually released, he was under constant observation during his stay in England.

Rickenbacker states he maintained his good nature throughout this episode, being afforded some small opportunities as a result. He managed to travel to several locations, including a nearby aerodrome. Apparently that visit in particular had a dramatic effect on his future because he became deeply fascinated with aviation. While sailing home he worked out plans for an air squadron to be composed of race car drivers, called "The Aero Reserves of America." A short time later, he approached the Army Signal Corps with the concept. However, they indicated race car drivers would not make good pilots, adding "...it would not be wise for a pilot to have any knowledge of engines and mechanics. Airplane engines are always breaking down, and a man who knew a great deal about engines would know if his engine wasn't functioning correctly and would be hesitant about going into combat."[3]

Rickenbacker then received a further disappointment; he was turned down in his initial efforts to enlist as a pilot. At 27 he was considered too old, and his education was believed rudimentary at best. Nevertheless, he was determined to enter the war. And on May 25, 1917, Rickenbacker joined the Signal Enlisted Reserve Corps, Aviation Section, with the rank of sergeant. Three days later he embarked for France with General Pershing's staff for assignment to AEF Aviation Headquarters in Paris. Although the press stated he was General Pershing's chauffeur, in truth, Rickenbacker "never did drive for the General."[4] He was, however, a chauffeur for Colonel Milling at the AEF Paris Headquarters, and this role brought him in contact with many officers, including Colonel William "Billy" Mitchell.

Rickenbacker standing next to a Cauldron G.3 trainer at the Aviation Training School at Tours, France, circa October 1917. His primary flight training lasted 17 days, with 25 hours flight time. (Photo George Williams.)

Raoul Lufbery standing beside a Nieuport 28. He obtained 17 victories while flying with the *Lafayette Escadrille* – the first American volunteer squadron to fly in France during World War I. In January 1918 Lufbery was transferred to the American Air Service with the rank of major. He was attached to the 94th Aero Squadron in February 1918 and was shot down in flames while attacking a German two-seater on May 19.

After much persistence and some luck, Rickenbacker was assigned to the Aviation Training School at Tours. This was illustrated by Rickenbacker when he stated: "The doctor who gave me the physical turned out to be another old friend. He pronounced me fit but also wrote my age down, firmly, as 25, with my birth date as October 8, 1892. My true age, 27, would have disqualified me."[5]

Rickenbacker was in the first group of American cadets sent to Tours, and one of the first to graduate. He described the speed of the program when he later wrote: "Our course of primary flying lasted exactly seventeen days… I put in a total of 25 hours of flying time and went forth a pilot and a first lieutenant in the Signal Corps."[6]

Rickenbacker completed the course at Tours on October 10, 1917, and was then sent to the 3rd Aviation Instruction Center (A.I.C.) at Issoudun, France. Issoudun was a flat, open tract of land far enough from Paris to be free of its worldly pleasures and large enough to provide room for flight training.

The U.S. Army had taken over Issoudun in September 1917. When Rickenbacker arrived in November there were still no facilities, and for the remainder of the year troops were busy setting up aircraft hangars, maintenance shops, offices, and living quarters.

Rickenbacker served as Engineering Officer, HQ Detachment; first under Captain James E. Miller, then under Major Carl Spaatz (who replaced Miller in December 1917). Rickenbacker would later state: "My duty was to organize the mechanical end of the school – repair shops, spare parts and other material, buildings, and transportation."[7] Yet, it was during this period that he obtained his advanced training on Nieuports.

In early January 1918, Rickenbacker requested transfer to a combat squadron. This was approved and he was assigned to the aerial gunnery school at Cazeaux, France, for advanced training.

In a letter from First Lieutenant Douglas Campbell (then a student at Cazeaux), a description of the base and its curriculum was given: "We are living at this very comfortable but reasonable hotel, and take a truck at 6:30

1/Lt. Douglas Campbell standing beside 1/Lt. Wentworth's Nieuport 28, serial N6168 "5". Campbell claimed the squadron's first victory, shooting down *Vzfw.* Wronicki from *Jasta 64w* on April 14, 1918. Although Wronicki's Pfalz D.IIIa crashed in flames, he was able to escape serious injury, being taken POW. Campbell eventually became the first American trained pursuit pilot to obtain "Ace" status, finishing the war with six victories.

1/Lt. Alan Winslow standing beside Nieuport 28, serial N6168 "5". This was 1/Lt. Wentworth's plane, decorated with a diamond backed rattlesnake draped over the cowling and fuselage. Winslow enlisted with the French Aviation Service on July 10, 1917, training at Pau. He first flew with French *Escadrille SPA 152*, from December 24, 1917 to February 12, 1918. He then transferred to the U.S. Air Service, commissioned a 2/Lt. and assigned to the 94th Aero Squadron. Winslow obtained his first victory (the squadron's second) on April 14, 1918. On July 31, 1918, he was shot down with a serious wound, his left arm being shattered by an explosive bullet. Winslow managed to land behind German lines before loosing consciousness from shock and loss of blood. However, the severity of the wound made it necessary to amputate the arm above the elbow. After five months in German hospitals, he was repatriated to France.

each morning to the school, which is about ten miles away on the shore of Cazeaux Lake (Etaing de Cazeaux). After eating French breakfast there, we are met by our special instructor, who conducts us to whatever lectures or practical work may be on the schedule.

"...Cazeaux is a beautiful place to fly. As soon as you climbed a couple of hundred meters, you can see two lakes, the estuary at Arcachon, and several miles of Atlantic coast. The only trouble is that there is no landing field in sight except the aerodrome itself, so if you have a 'panne de moteur' when you are out of gliding distance, there is nothing to land on except pine trees. This would not be very pleasant, although it has been done many times without injury to the pilot...

"The machines we have been using are pretty punk, for they have seen a lot of service. It's good practice, though, for they are plenty strong enough, but the wings are warped a little, and they do all sorts of peculiar

things."(8)

Rickenbacker was among the first group of American pilots to complete the training program at Cazeaux. He was assigned to the 94th Aero Squadron, then being organized at Issoudun.

The 94th Aero Squadron had been officially created on August 20, 1917, at Camp Kelly, later Kelly Field, San Antonio, Texas. After a short training program in Texas, the squadron's ground personnel departed for Mineola

A view of the cockpit and front cowling area of the Nieuport 28 C1 with its machines guns removed. (Photo R.L. Cavanagh.)

Field in Garden City, New York, on September 30, 1917. Subsequent to the usual processing and waiting, the unit boarded the S.S. *Adriatic* for travel overseas, sailing on October 27. The 94th Aero reached England on November 10 and departed that same day for France, disembarking at the coastal town of Harve on November 13. The unit then traveled to Paris, there the ground personnel (mechanics) were split up into seven detachments, with each going to

Winslow (standing center) poses beside his first victory, an Albatros D.V, shot down on April 14, 1918. The plane's pilot, *Uffz.* Simon from *Jasta 64w,* was taken POW. From takeoff to landing, the Winslow/Campbell flight of April 14, 1918, lasted less than five minutes. (Photo George Williams.)

Rickenbacker seated in the cockpit of a Nieuport 28 (serial unknown). The light colored cowling (presumably white) indicates this is an aircraft from the First Flight. However, this is not Rickenbacker's "White 12" (N6159). The unit insignia reveals slight differences when compared to the insignia on the "White 12" machines – this insignia's ring crosses over two red hat band, whereas the ring on "White 12" (N6159) crosses over one red hat band. In addition, the insignia's placement over the fuselage ribs reveals location differences.

a separate aircraft facility (Breguet, Brazier, Renault, Nieuport, Gnome, Blériot, and Hispano-Suiza) for a ten-week training session.

In February 1918 pilots and mechanics of the 94th Aero Squadron were reassigned to Issoudun. There the unit was placed under the command of Major Jean W.F.M. Huffer, a flyer of considerable reputation, having served with the *Lafayette Escadrille*.[9] Soon after, the unit moved to Villeneuve-Les-Vertus, arriving by train on March 6, 1918. Yet, valuable time would pass before aircraft – Nieuport 28s – were made available to them. In fact, at that time the machines were still being collected from Villacoublay and Colombey. But in an attempt to accelerate the process, seven aircraft were eventually taken from the 95th Aero Squadron after that unit was pulled from the front and reassigned to Cazeaux for training.

The first Nieuports to reach the 94th Aero arrived on March 18. That same day ground crews assembled and readied these aircraft for operations, except for machine guns, which were still not available. Undaunted by lack of weapons, the following day, March 19, the unit conducted

its first (unarmed) patrol over enemy lines, flying between Rheims and the Argonne Forest. The patrol consisted of Major Lufbery and First Lieutenants Rickenbacker and Campbell.

Campbell described the event in a letter: "The morning was beautifully clear, and after a leisurely breakfast we crawled into our warm duds and took to the air, at about 9 o'clock."

"Major Lufbery led, with Rick [Rickenbacker] and myself above and behind him on the right and left, respectively. With these fast and powerful little machines it didn't take us long to reach the lines at an altitude of 4,000 meters. Then we turned to the left and patrolled up and down two times on a front of about 30 kilometers. Our planes were scarcely ever flying in a straight line; we were continually banking and turning from one side to the other in order to see at all angles and to make it harder for the anti-aircraft battery which shot at us occasionally."[10]

Although Rickenbacker was concentrating on fighting off a serious case of airsickness, his comments better illustrate the accurate anti-aircraft fire: "I felt the cold

Nieuport N6159 "12" with Rickenbacker standing along side. He obtained victories one and two in this machine. The edge of the Liberty Loan poster is just visible on the top right wing panel.

sweat break out all over me and the acute misery of nausea well up from my stomach. I was on the verge of disgracing myself, when suddenly the plane rocked violently and a burst of light and sound hit my eyes and ears. Another blast rocked me and another and another. I look behind me. Large puffs of black smoke marked my path through the sky. It was Archie [anti-aircraft fire]."[11] Rickenbacker stated: "I was startled by an explosion which seemed to crash out only a few feet behind me…for with the initial concussion my plane began to roll and toss much worse than before. The very terror within me drove away all thoughts of airsickness and in the next few minutes several more roars buffeted my plane and the repeated thuds of continued explosions hammered my ears."[12]

Fortunately, the Nieuports flew quickly beyond the range of anti-aircraft fire and continued their patrol without further incident. When landing at the aerodrome some time later, the three men were surround by a large group who were apparently eager to hear details of the squadron's first mission into enemy territory.

By Rickenbacker's account, he and Campbell were just as eager to detail the event. However, in an effort to appear more like Lufbery, the well-seasoned veteran, the two young lieutenants wore faces of "bored indifference," while nonchalantly stating "We had had a little flip-around over the Hun [German] batteries and it had been most droll seeing the gunners wasting their

1/Lt. Edward Rickenbacker, 1/Lt. Douglas Campbell, and Captain Kenneth Marr (squadron C.O.) standing next to a Nieuport 28 (serial unknown). This is the same Nieuport shown on page 56. As stated, the light colored cowling (presumably white) indicates this aircraft was from the First Flight. However, it appears the landing gear struts and center portion of the propeller have been painted a light color too (presumably white again). The dark streaks appearing on the landing gear struts is presumably paint from the French camouflage pattern showing through. (Photo George Williams.)

Captain James Hall, standing besides Nieuport 28 N6153 "17", circa April 1918. A single gun is mounted to the aircraft. Hall shared a victory with Rickenbacker on April 29, 1918. However, during the battle of May 7, 1918, he was forced down in altitude, where a German anti-aircraft shell hit his engine. He crash-landed with minor injuries, spending the rest of the war a POW. After the war, Hall teamed with Charles Nordhoff to write *The Lafayette Flying Corps,* the official history of the unit, and numerous best sellers including the classic *Mutiny on the Bounty.*

ammunition."(13) When the two men were initially questioned on enemy aircraft, Rickenbacker added "...none of them dared to venture up against us. Not a plane was in our vicinity."(14) Apparently at this point Lufbery jumped into the conversation and began to discuss the many French and German aircraft that had crossed their patrol's flight path.

Rickenbacker later admitted that "Lufbery paused and looked at us as though we were inattentive children. Then he grinned and walked over to my plane. He poked his finger through a hole in the tail and another through the wing; then he pointed to where another piece of shrapnel had gone through both wings not a foot from the cockpit."(15)

Later that night Campbell penned a letter home. It read in part: "...The Major [Lufbery] saw one *Boche* [German] flying very low, but my untrained eye didn't even observe that...I guess we will next time, and anyway it is something to have been one of the first two American-trained *chasse* [sic] pilots over the lines."(16)

Getting over the lines was paramount in everyone's mind, because a majority of pilots from the 94th Aero were

in desperate need of combat experience and confidence. Addressing the issue, French aviators were called in to escort the Americans on daily patrols. However, these activities were stopped on March 21, 1918, after the Germans launched a series of "drives" that assaulted the Allied front from the Channel to Rheims. In response, every available British and French air unit was pressed into the defense. Meanwhile, the 94th Aero was still unarmed and forced to remain out of harm's way, momentarily at Villeneuve-Les-Vertus, but being pulled back to Epeiz on March 30, as the Germans continued to advance.

The unit's first shipment of machine guns finally arrived in early April 1918. Despite the fact quantities allowed only one gun per aircraft instead of the required two, the unit was now ready to commence combat operations. As part of their combat indoctrination, the 94th Aero Squadron was transferred to the quieter Toul sector of Lorraine, being attached to the French Eighth Army and stationed at the Gengoult airfield on April 11.

The Toul sector was justly considered a quieter section of the front; its front lines had changed very little in the three and a half years of fighting, and both the French and

First Lieutenant Edwin "Eddie" Green poses with his Nieuport 28, N6180 "18". Green was involved in the dogfight of May 7, 1918.

A lineup of Nieuport 28s during the award ceremonies at Gengoult on May 15, 1918. All the aircraft have light colored cowlings. Among those Nieuport 28s visible in the lineup is N6185 "10", often incorrectly identified as 1/Lt. Campbell's plane. In fact, it was the spare plane of the First Flight, being flown by 1/Lt. Gude during the combat in which Major Lufbery was killed. Nieuport N6230 "4" was forced landed by 1/Lt. Smythe on May 23. Nieuport N6154 "7" was Gude's assigned aircraft. Nieuport N6159 "12", flown by Rickenbacker, displays the Liberty Loan poster on the upper right wing panel. (Photo George Williams.)

Nieuport N6159 "12". The serial number N6159 appears in small black characters on rudder's white colored panel. Also visible is the lower right wing's Liberty Loan poster and Rickenbacker's medallion, mounted near the cockpit's starboard side. (Photo Greg VanWyngarden.)

The white-cowled Nieuport 28s of Rickenbacker's flight. Second aircraft from left is N6159 "12", behind which can be seen Nieuport N6169 "1", the regular machine of Major John Huffer, but Rickenbacker would score four victories in it.

Germans tended to use the region as a rest zone. For these same reasons the region was selected for the Americans in order to give them time to gain experience before being transferred to Chateau Thierry, where the main German offensive was going on, and where first-class opposition awaited them.

On April 14, 1918, the 94th began active patrols over the sector of St. Mihiel to Pont-a-Mousson. That same day, First Lieutenants Campbell and Winslow obtained the unit's first victories after taking off from the Gengoult airfield to answer an alert. Four and one-half minutes later, the Americans had shot down two German fighters from *Jasta 64w* – an Albatros D.Va and a Pfalz D.IIIa.

On April 29, 1918, the 94th had another run-in with *Jasta 64w* over St. Baussants. In the ensuing battle, a Pfalz was forced down and credited as a shared victory between Captain James Hall (third victory) and Rickenbacker (first victory). Their joint combat report stated: "…attacked Albatros monoplace at about 6:10p.m. between Baussane and Montsec. Capt. Hall fired, and also Lt. Rickenbacker, 200 rounds. Enemy machine dove, Capt. Hall and Lt. Rickenbacker followed him, firing all the time. Cloud of smoke emerged from enemy machine and continued to pour out all the way down. Lost sight of enemy at 700 meters near Vigneuilles-les-Hatten Chatel, still going down. Redressed at 700 meters. Confirmation asked."

Nieuport N6169 "1" with Rickenbacker on deck. The aircraft's cowling was painted in USAS roundel colors of white/blue/ red (fore to aft). In addition, the white numeral "1" that appeared on each side of the fuselage was further decorated with shadowing in red and blue. (Photo Harold H. Tittmann collection; NASM via Alan D. Toelle.)

Rickenbacker standing next to Nieuport 28 serial N6169 "1" after a forced landing on May 17, 1918. Most of the top wing's fabric pealed off as he pulled the aircraft out of a dive. Although the plane retained some level of controllability, Rickenbacker was lucky to bring it down for a forced landing. His facial expression reveals his anxiety. That night squadron ground crews repaired damage sustained in the landing and replaced its wings; the next day N6169 "1" was ready for operations. Rickenbacker would score three more victories in it. (Photo George Williams.)

Nieuport N6169 "1" at the gun range. Although originally assigned to Major Huffer (squadron C.O.), Rickenbacker often used it, scoring victories 3–6 in N6169 during May 1918. (Photo Harold H. Tittmann collection; NASM via Alan D. Toelle.)

Nieuport N6283 "16" photographed at Gengoult, France, circa late June 1918. This aircraft was assigned to Rickenbacker from May 29 to June 7, 1918. It was then transferred to 1/Lt. Smythe. This aircraft's rudder illustrates the factory-applied American markings. The upper right wing also bears the diagonal squadron identification stripes. In the background is 1/Lt. Tittmann's plane, N6271 "7", shown prior to the squadron insignia being applied. (Photo M. Edwin Green album via Alan D. Toelle.)

The following day, a confirmation report was filed by First Lieutenant A. Fredrick Oberlin, 102nd U.S. Infantry: "Two of our planes appeared over Beaumont at about 6:00p.m. and circled above the German lines in the vicinity of St. Bauxxant [Baussane]. A few moments later a third plane appeared on the scene. All planes were at a fairly high altitude. The two American planes separated, one going toward the W [west] the other maintaining his position.

"Machine gun fire was then exchanged between the German and American plane. Our plane seemed to climb above his adversary and stated to dive using his M.G., whereupon the German turned and dove for the earth followed by our aviator. The German landed so far behind the lines that it was impossible to see just where, as he was lost in the haze that was quite heavy at that time.

"The movement that drove the German to the ground was nicely executed, our aviator leading him on and at the same time climbing above him and turning suddenly and diving toward him. Plane was driven down in the direction of ESSEX."

On May 4, 1918, the 94th and 95th Aero Squadrons

Members of the 94th Aero Squadron, at Toul, France, circa June 1918. Standing left to right: Thorne Taylor, "Bill" Loomis, "Jimmy" Meissner, Edward Rickenbacker, and "Walt" Smyth. Loomis was from originally from SPA 153 and later went to 213th Pursuit Squadron. Meissner was later made C.O. of the 147th squadron. Smyth was killed in action in August 1918.

Rickenbacker with his Kellner-built SPAD 13, serial S4523 "1", circa August 1918. The large numeral "1" is visible on the undersurface of the lower left wing (presumably white, outlined in red). Likewise, the squadron's diagonal red and white stripe can be seen on the undersurface of the lower right wing. Additional markings included a red cowling and tri-colored bands – red, white, and blue – which wrapped completely around the forward landing gear struts. (Photo George Williams.)

were reorganized into the American First Pursuit Group, the two squadrons occupying opposite ends of the Gengoult airfield.

On May 7 Captain Hall and First Lieutenants Rickenbacker and Green spotted three enemy aircraft above Pont-a-Mousson. The Germans, from *Jasta 64w*, were patrolling over the sector in an effort to guard the King of Saxony while he reviewed the troops. Apparently they were unaware of the Americans diving down from 2,500 meter until the two groups met at 8:05a.m. In the resulting combat, at least one bullet fired from Rickenbacker's guns hit *Leutnant* Willi Scheerer in the stomach. Although he escaped the battle, he crash-landed and a few hours later died of his wound. Meanwhile, *Leutnant* Friedrich Hengst got on Captain Hall's tail and drove him down in altitude. Moments later an anti-aircraft shell fired from a flak unit (*M.Flak 54*) hit his engine, forcing him to crash-land. Yet Hall's injuries were minor and he spent the rest of the war a POW.[17]

Rickenbacker's initial combat report read: "While flying over Pont-a-Mousson. Saw 3 enemy planes about 5 kms in and about 2,500 meters high. I started to attack, being followed by Capt. Hall and Lt. Green. The fight took place about 8:05a.m. Last saw of Capt. Hall he was in a *vrille*[18] about 1,500 meters high northwest of Pont-a-Mousson, 5 kms in Germany. The enemy plane that I attacked was last seen diving home. I fired about 50 rounds then met Lt. Green and came home. Confirmation requested."

However, after Hall's loss was confirmed, Rickenbacker filed a more detailed account of the May 7 battle. It read: "Capt. Hall received an alert about 7:30a.m. of four enemy planes coming south Pont-a-Mousson. Captain Hall, Lieutenant Green and myself left immediately, flying Northeast, Pont-a-Mousson, arriving over Pont-a-Mousson shortly before 8 o'clock at about 3,500 meters altitude, but could not see anything of enemy planes. He then returned patrol east of Pont-a-Mousson for a short distance. Turning around, we again flew over Pont-a-Mousson at the same altitude, when I noticed one enemy biplace [two-seater] ahead of us just over the lines and about 1,000 meters below. However, their anti-aircraft give signal of our approach, he immediately diving into Germany. About the same time I noticed three enemy planes flying in formation approximately five or six kilometers in Germany and 1,500 meters below us. I attracted Captain Hall's attention and reversed directions, upon which we peeked towards the Boche. I, of course, was ahead of him at the time, owing to my position.

"I decided to attack one monoplace [single-seater] and

Rickenbacker seated in SPAD S4523 "1", circa August 1918. His medallion has been mounted near the cockpit's starboard side. (Photo George Williams.)

was peeking close to him when another appeared from the rear. I left the first one, started climbing, and again peeked on another, firing approximately 50 or 75 rounds at him. The last seen, he was still peeking into Germany. While in a village, I noticed a Nieuport going down in a vrille. Of course, at the time I was unaware who it was. I then saw one of the patrol headed back towards our lines and immediately started climbing in the same direction, as the enemy had been so scattered it was useless to continue the fight. Upon arriving at the field, I found that Captain Hall was missing."

Shortly after his second victory, Rickenbacker was put in charge of the squadron's First Flight. In the week that followed, he obtained a third victory and had a narrow escape with death. Both events occurred during the same patrol on May 17 while he dove through a formation of three German fighters "at a furious pace."[19]

Rickenbacker's fire hit an Albatros D.V in the maneuver. However, he admitted he had prolonged the dive's "terrific speed a trifle too long,"[20] and that he had "come out of it in a hurry."[21] As a result, the force of the air flow over the aircraft was too great, and most of the fabric pealed off the top wing at an altitude of 4,500 meters. Initially Rickenbacker "manipulated the controls, but it did no good. The plane turned over on her right side. The tail was forced up. The left wing came around. The

ship was in a tailspin. With the nose down, the tail began revolving to the right, faster and faster."[22] Luckily, the Nieuport 28's ailerons were on the lower wing, and he was able to regain control. With only 1,000 meters altitude remaining, he leveled the aircraft and then brought it down for a forced landing.

Rickenbacker's combat report was concise: "Had combat with three Albatroses in vicinity of Richecourt at 6:24. One Albatros was last seen descending in a vrille, the other two going into Germany. One plane evidently landed just behind German line, badly damaged. Confirmation Asked." However, it seems strange he made no mention the fabric pealing off the top wing.

The Nieuport 28s were known to have problems concerning wing fabric failure. But just as serious, and perhaps not as well known, N.28s were prone to engine vibrations that caused cracking in the aircraft's rigid copper fuel lines. The result was frequently an explosive fire that killed the aircraft's pilot. This was illustrated on May 22, 1918, during a voluntary patrol involving First Lieutenants Rickenbacker, Chambers, and Kurtz. Moments into the flight, the three Americans engaged three German fighters north of Flirey. Rickenbacker was credited with an Albatros D.V (his fourth victory), while the Americans escaped unharmed. Yet, while preparing to land at the Gengoult airfield, Kurtz was killed when his Nieuport 28

Rickenbacker with his Kellner-built SPAD 13, serial S4523 "1", circa August 1918. He obtained victories 7–26 in this machine. The outer wheel covers were blue and contained a white star with red dot. Also visible is the top wing's white numeral "1" with a narrow dark outline (presumably red). The rudder is finished in blue, white, red stripes (forward to aft). The rudder's text format is unique to Kellner, yet it appears some rework has been done.

burst into flames and crashed in the nearby trenches.

Rickenbacker's combat report read: "Saw three Monoplace Albatros 4,500 meters over St. Mihiel at 9 h 12 approximately. Attacked and followed them into Germany about six kilometers, separating one from the group, which I again attacked and which fell, shooting about 100 rounds. Enemy plane last seen diving steeply North of Flirey."

The following week, on May 28, Rickenbacker teamed up with Campbell for a patrol over the region of St. Banssant. The two Americans attacked an enemy two-seater that spun down over Bois Rate. But as no loss was reported, the crew seems to have survived unharmed. Rickenbacker's combat report read: "First noticed six enemy planes coming south at 9 h 25, 2 Albatros biplace and 4 monoplace Albatros. Attacked 1 biplace Albatros at 9 h 31. After firing 75 rounds saw him go into vrille and watched him crash in Bois Rate at 9 h 35 attacked another biplace in region of St. Banssant, altitude 2,500 meters. Fired about 100 rounds. Last saw him going into vrille, but could not see him crash. Then returned to our lines account no gasoline."

On May 30, Rickenbacker scored his sixth victory, claiming a two-seater over Jaulny. Fifteen minutes later, 1/Lt. Jimmy Meissner entered a dogfight with a German fighter east of Thiaucourt, during which the German's landing gear tore into Meissner's upper wing, resulting in fabric loss. Rickenbacker intervened against Meissner's attackers, allowing him to bring his damaged plane safely

down.

Rickenbacker's combat report read: "Had one combat at 4,000 meters over Jaulney about 7.38a.m. Five German planes two of which I thought were Albatroses Biplace and three Albatros monoplace. Fired about 100 rounds and a biplace fell into a vrille after a few moments combat. Attacked another monoplace a few minutes later, who piqued back into Germany. I then left the fight as the enemy planes were too low and too far into continue. Returned to our lines and noticed four more planes coming toward the lines. I returned to the attack and found several Nieuports engaged in same, having arrived shortly before I did. Noticed Lt. Meissner piquing on one Boche and another piquing on him, at whom I fired about 50 rounds. During his maneuvers, he and Lt. Meissner collided, Lt. Meissner losing most of his upper wings. Started for our lines with a biplace taking note of his damaged condition. Piqued on biplace and fired about 50 rounds. He then left the combat, Lt. Meissner and I returning to the camp. This combat took place about 7:55a.m. east of Thiacourt. There were two Nieuports engaging in combat with the balance of the Germans when I left. Confirmation Requested."

Throughout June 1918, Rickenbacker engaged several enemy aircraft, but none of the combats were decisive. Then on June 27, the unit was shipped out, along with the other three squadrons of the First Pursuit Group, to the Touquin aerodrome in the hotly contested Chateau Thierry sector.

Starboard side of Rickenbacker's SPAD S4523 "1", circa September 1918. A fourth battle damage patch (white circle) can be seen on the lower right wing's trailing edge. The top wing displays a white numeral "1" on the right panel and the squadron's diagonal stripe, of red and white, on the left. The aircraft's snubbed end exhaust stack is clearly visible in this photo.

Rickenbacker would fly only two patrols in July 1918, before sharp pains in his right ear caused him to be invalided to Paris.[23] There he learned that he had a severe abscess that had to be lanced and treated. He would not return to the squadron until July 31, and would fly only irregular missions until September 1918.[24]

In preparation for the next Allied offensive, on September 1, 1918, the First Pursuit Group was transferred to an airfield at Rembercourt on the St. Mihiel sector, arriving on September 2 and conducting its first patrol the next day. On September 12 the St. Mihiel Offensive began with the First Pursuit Group launching missions at daybreak. The enemy seemed very active on this front and the patrols were continually meeting low-flying reconnaissance machines protected by large formations of Fokker D.VIIs.

On September 14 Rickenbacker scored his first victory in a SPAD 13 and his seventh victory of the war. He filed the following report: "Left at 7 h 30a.m. met one enemy biplace over Verdun. Looked like an L.V.G. Followed him about seven (7) Kilometers, but it was too high to reach, Then returned from the lines at 4,000 meters. Met four (4) enemy Fokker with red wings, light gray fuselage and striped tail over the towns of Villency and Wayville. They were flying at about 3,000 meters. I piqued on the upper man of the formation, fired approximately 200 rounds and saw him go down, apparently out of control. Was unable to follow him on account of the other three, who showed excellent fighting spirits. This took place at 8 h 10 and 8 h 15. I then returned over Lachausee, where, at 8 h 25, I

noticed five (5) Fokkers with their regular camouflage crosses on the tail, apparently new men. I started to pique, but they immediately turned for home. Fired about 50 rounds without result. Confirmation Requested."

On September 15 Rickenbacker downed another Fokker D.VII, this time over Bois de Warville. His combat report stated: "Encountered six (6) E.A. with regular camouflage at 8 h 10 in the region of Bois de Waiville and the Bois de Bavonville after having flown up and down the sector with them parallel with the lines at about the same altitude. Then returning over the above mentioned region this patrol piqued on four SPADs, one of which I noticed to go down in a vrille. I saw my opportunity to pique on the last man of the E.A. patrol, which was a Fokker type. I got several good bursts into the cockpit and noticed the E.A. to go down flopping from side to side out of control. The encounter started at 4,500 meters and ended at 2,000 meters. At the time that I piqued I noticed nine (9) E.A. flying from the direction of Metz over Arz sur Mosell. Confirmation Requested."

September 25, 1918, was an eventful day for Rickenbacker. He was promoted Captain, replacing Captain Marr as Commanding officer of the 94th Aero Squadron. And later, during a morning patrol over Billy, he single-handedly engaged seven enemy aircraft, downing two for his first double victory of the war. He was eventually awarded the Medal of Honor for this act of bravery.

By his own account, Rickenbacker's attack was swift. He was well above the enemy formation when he stated "I

Rickenbacker with SPAD 13, serial S4523 "1", circa September 24, 1918. Visible are three battle damage patches (white circles) on fin and fuselage. However, the small black crosses have yet to be painted in the patches. Compared to the earlier photos of S4523, this photo reveals an exhaust stack of different shape; it has a snubbed end. The rudder is finished in red, white, and blue stripes (forward to aft). The rudder's text format is unique to Kellner, yet it is different from the format that appeared on S4523 in earlier photos.

shut down my engine, put down my nose and made a bee line for the nearest Fokker. I was not observed by the enemy until it was too late for him to escape. I had him exactly in my sights when I pressed both triggers for a long burst. He made a sudden attempt to pull away, but my bullets were already ripping through his fuselage and he must have been killed instantly. His machine fell away and crashed…"[25]

Still diving, Rickenbacker "plunged straight on through their formation" to attack the photographic machines. However, he acknowledged "the victory was not to be an easy one. The pilot suddenly kicked his tail around, giving the gunner another good shot at me. I had to postpone shooting until I had more time for my own aiming And in the meantime the second photographing machine had stolen up behind me and I saw tracer bullets go whizzing and streaking past my face. I zoomed diagonally out of range, made a reversement[26] and came directly back at my first target,"[27] adding "They were flying parallel to each other and not 50 yards apart. Dropping down in a sideslip until I had one machine between me and the other, I straightened out smartly, leveled my SPAD and began firing. The nearest Boche [German] passed directly through my line of fire and just as I ceased firing I had the satisfaction of seeing him burst

into flames."[28]

His combat report read: "Patrolled the line from Verdun to the Lake Lachaussee. At 8 h 15 I noticed twelve (12) monoplace Fokkers in the region of Mars-la-Tour. After staying on the line for about ten (10) minutes, they returned toward Metz. At 8 h 40 I noticed our anti-aircraft shells bursting in the region of Billy. Saw five (5) Fokkers protecting two (2) biplace planes coming toward our lines. I was flying 500 meters and they were approximately 3,000 meters. Dove out of the sun onto rear man. The last seen he was going down out of control. Fired about 100 rounds at close range. Continued my dive through the formation and fired about 100 rounds at a Halberstadt biplace from about 50 to 70 meters. This combat lasted about three minutes. When last seen, from an altitude of 200 meters, he was going down in a very steep nose dive. His protection then returned from a greater altitude then I had, forcing me to retire. Two Confirmations requested."

The next day, September 26, Rickenbacker obtained his 11th victory, shooting down a Fokker D.VII over Damviller. By Rickenbacker's account the German pilot "was looking for a fight. He gunned his plane, went out ahead, banked around and came at me head on. We both started firing at the same time…coming together at a combined speed of four miles a minute. I thought that he

Standing left to right: Joseph Eastman, James "Jimmy" Meissner, Edward Rickenbacker, Reed Chambers, and Thorne Taylor. A fifth bullet damage patch is visible on the lower left wing, near the fuselage. The fin's two battle damage patches now have their small black crosses. At far right is the Hannover CL.IIIa that Rickenbacker downed for his 14th victory.

had better get the hell out my way. I certainly was not going to get out of his. Just as we were about to crash head on, he dived under my plane. I immediately put my SPAD into a reversement – pulled the stick straight back to start a loop and simultaneously rolled in over in a half turn. I came over on his tail right side up and ready to shoot. I gave him a long burst of bullets. He began the long fall down."(29) However, Rickenbacker's SPAD had also sustained serious damage in the battle. Within seconds, its engine began to vibrate so badly that it "almost tore itself out of the frame."(30) Being over German-held territory, he was lucky to nurse the aircraft back into Allied lines. In addition to numerous bullet holes, once Rickenbacker landed, he discovered that half of the SPAD's propeller blade was gone, shot off during the battle. His combat report states: "Left ground at 5 h 20 o'clock. Arrived at the enemy balloons in our sector at about 12 minutes of six. Was unable to locate either one, so I flew east about five kilometers in their lines. Noticed one balloon go down in the region of Damvillers at approximately seven minutes of six o'clock. This balloon, I found out later, was shot down by Lieut. Vasconcellas of the 27th Aero Squadron. At the same time I noticed what I thought to be another

balloon go down in the region of Montefancon. Continued to fly in the region of Damvillers, when I met up with a Fokker at 1,500 meters, at whom I fired about 100 rounds and noticed him go down apparently out of control. During the combat my propeller was shot up badly, forcing me to retire and land on the auxiliary field at Verdun. Was just about to land when I noticed a large flare in the region of Stain but, was unable to state what it was. After changing propellers and refilling, I returned to the lines. Flew up to the Argonne Wood, back over the Fresnes and only noticed one enemy balloon and that to the south of the Lachaussee. Confirmation requested on Fokker plane."

Two days later, September 28, Rickenbacker claimed his first balloon during an early morning strafing mission. His report stated: "Left field at 5 h 20. Followed Meuse River up to Milly about 5 kilometers, on the west side, then I turned down the main road on the east side of river at an altitude of 200 to 300 meters. Then I noticed balloon being hauled directly toward me by a truck between Vilosnes-sur Meuse and Sivry-sur-Meuse. It was about 100 meters high. Fired about 150 rounds and noticed it go down in Flames between 6 h 00 and 7 h 05. I was fired on by M.G., one

SPAD S4523 "1" with Rickenbacker on deck. The diagonal stripe (red and white) can be seen on the upper wing's leading edge. The cowling color (presumably red) also shows on the front portion of the upper engine cowling. The aircraft's twin 0.303 Vickers machine guns each fired at a rate of 850 rounds per minute.

bullet reaching my main strut in my tail. Sounded so loud I landed at Verdun for inspection and obtained from there another supply of gas. Confirmation Requested."

Another enemy balloon followed on October 1, 1918. Rickenbacker reported: "Saw enemy balloon on the ground in the vicinity of Puxieux. Piqued on it and fired about 75 rounds into it. It went up in flames. It would appear from an observers point of view exactly like an ammunition dump explosion on account of being on the ground. Confirmations Requested."

The following day, Rickenbacker and 1/Lt. Chambers downed Hannover CL.III serial 2392/18 over Montfaucon at 5:30 p.m. but did not realize a the time that they had downed it.[31] Being determined, ten minutes later Rickenbacker engaged a large group of Fokker D.VIIs over Vilosnes and shot one down for his 15th victory. His combat report reads: "Weather Poor. Altitude 1,000 meters. Met one Halberstadt north of Mount Faucon at 17 h 30 Fired about 100 rounds. No apparent results. About 17 h 40 met 7 or 8 Fokkers at 1,000 meters in region of Vilosnes. Fired about 250 rounds when I noticed one Fokker go down out of control. Was forced to leave home on account

of the enemy's advantage. Confirmation Requested."

On October 3, Rickenbacker scored another double, being credited with a Rumpler over Clery-le-Grand at 4:40p.m. and a L.V.G. two-seater over Dannevaux at 5:07p.m. His combat report reads: "Was alone ahead of our protection at 16 h 37, saw balloon go down in flames. Attacked (1) biplace Rumpler just over Clery-le-Grand at 16 h 40. Fired about 100 rounds when he was seen to go down by Lieut. Crocker (94th). Returned to our lines, was patrolling same when I noticed two (2) L.V.G. planes in our lines. Attacked one (1), fired about 150 rounds when he went down in flames. Lieut. Coolidge and Curtis (95th) also attacked the same plane and should share in the victory. This plane fell in flames just west of Dannevoux. Confirmations requested: 1 Rumpler and 1 L.V.G."

The following week, on October 9, Rickenbacker was conducting a voluntary patrol between Dun and Marveux when he burned another balloon at 600 meters for his18th victory. His combat report stated: "Saw two (2) Reglage machines on their side of the line. I returned to Montfaucon until they disappeared, which about 17 h 25. Then started up the river at 17 h 30. A few kilometers

Above & below: Hannover CL.IIIa 3892/18 shot down near Montfaucon, France, by Rickenbacker and 1/Lt. Reed Chambers on Oct. 2, 1918. Apparently this victory came as a surprise to Rickenbacker; his combat report simply stated: "…Fired about 100 rounds. No apparent results." It also incorrectly called the German aircraft a "Halberstadt." Nevertheless, the machine crashed landed in Allied lines for Rickenbacker's 14th victory. The Hannover was a tough, maneuverable two-seat fighter.

Right: Rickenbacker with his SPAD. This photo indicates the star design was not carried on the inside of the wheel covers.

Below right: Another view of Rickenbacker with his SPAD. This photo shows a 7th battle damage patch near the left wing root.

Below: Rickenbacker with his SPAD showing the hat-in-the-ring insignia of the 94th Aero Squadron.

Close-up of the squadron insignia located on the starboard side of Rickenbacker's SPAD 13, S4523 "1". This post-war photograph shows 26 crosses, representing the number of victories Rickenbacker obtained during the war. The aircraft's seventh battle damage patch is visible with small black cross added. (Photo George Williams.)

beyond Don, returning found the balloon on the edge of the woods on the ground. Attacked it about 17 h 52. I fired about 150 rounds when it caught fire, making quite a blaze. I then returned to the airdrome at 18 h 05. Confirmation requested – one (1) Balloon."

On October 10, Rickenbacker scored another double, claiming two Fokker D.VIIs over Clery-le-Petit at 3:52p.m. At least one of the Germans – Lt. Kohlbach, from *Jasta 10,* parachuted to safety. He later report: "Followed formation over Doulcon when five (5) Fokkers came down on the formation from the southeast. I Piqued on one in the region of Clery-le-Petit and fired about 150 rounds at close range when I was forced to dive on account of other Fokkers behind. Did not follow him down. This combat took place about 15 h 52 at which time Lieut. Chambers claims a Fokker fell in flames. I then noticed a SPAD below being chased by a Fokker. I piqued on him and fired about 100 rounds into him. When last seen, he was in a steep bank. Confirmation requested – Two (2) Fokkers."

Rickenbacker downed another Fokker D.VII over Clery-le-Petit on October 22 for his 21st victory. His combat report read: "Went up the east side of the Meuse River at 1,200 meters. I noticed a Fokker pique on a Breguet, and I piqued on the Fokker, firing about 100 rounds into him at close range. This was in the region half way between Clery-le-Petit and Dun-sur-Meuse at about 15 h 55. When I last saw him he was going down out of control at about 700 meters. Lt. Maye reports seeing a plane go down in flames from about 400 meters at 15 h 55 in the region north of Liny. About 16 h 00 I noticed another Fokker pique on a Salmson just north of Cunel. Fired about 50 rounds at long range. When last seen he was piquing into Germany. Lieut. Chambers also fired at this same Fokker and the E.A. did not regain his formation. Confirmation Requested – One (1) Fokker."

The next day Rickenbacker engaged *Vzfw.* Gustav Klaudat from *Jasta 15.* In the ensuing battle Klaudat was wounded when a bullet shattered the bone of his left upper arm. Rickenbacker's combat report stated; "While

patrolling just north of Montfaucon I noticed one (1) of our balloons going down in flames west of Montfaucon. Saw the Fokker who attacked the balloon headed for the enemy lines. Started to cut him off and while watching him ran into four Fokkers, probably his protection. Fired about 100 rounds into one of them at close range. This took place at about 16 h 55 in the region of Le Grande Carre Ferme. When last seen the Fokker was in a steep nose dive. Was unable to follow him any further as the three others began firing on me. Altitude of combat 600 meters. Confirmation Requested – one (1) Fokker."

One week later, on October 27, Rickenbacker claimed another pair of Fokkers D.VIIs. One was piloted by *Ltn.d.R.* Max Kliefoth from *Jasta 19,* who was shot down by Rickenbacker while he tried to engage a two-seater. Kliefoth survived the crash and was interned as a POW. Considering his aircraft was in captivity, he admitted *Jasta 19* had fuselage colors of yellow/blue. However, during the interrogation, Kliefoth try to lie about other units by stating that the "colors of JGII were, *Jasta 12:* black nose and yellow fuselage, *Jasta 12:* red nose and green fuselage, and *Jasta 15:* white nose and red fuselage."[32] Like any good prisoner, Kliefoth was giving false information to the enemy!

Rickenbacker's combat report for October 27 read: "While flying above our low patrol, was attacked by three (3) Fokkers from above. After a short dive I pulled up and turned on the end Fokker, firing about 125 rounds at close range. When last seen he was going down out of control, about two (2) kilometers northwest of Grand Pre. Time, about 14 h 50. Altitude, about 2,000 meters. At 15 h 05 in the region of Bois de Money, engaged one (1) Fokker firing at one of our bombers. Fired about 200 rounds and noticed his motor stop. I ceased firing and then started maneuvering him into our lines. Just before he landed another SPAD tried to shot him down. I placed myself between them and the Fokker landed safely in the region of the southern edge of Bois de Money, turning up on his nose. He had a blue fuselage with white circles on it and usual crosses on the wings. Altitude 3,000 meters to ground. Confirmation requested – Two (2) Fokkers."

On October 30, 1918, Rickenbacker obtained his 25th and 26th victories – his last of the war. His report stated: " Was following our formation when two (2) Fokkers with red fuselages and red noses piqued on rear end of formation. I fired about 50 rounds at long range without results. Went up the line towards the Meuse To Remonville and return within enemy lines about two (2) kilometers. In the region north of St. Juvin I met the same two (2) Fokkers. Fired about 100 rounds at one of them at close range before I was seen. He went down out of control; his gas tank exploded when he crashed on the ground. Other Fokker disappeared into Germany. At this time I was about 200 meters high. This took place about 16 h 35 in the region north of St. Juvin. Confirmation requested.

"About two minutes later I noticed a balloon in its bed directly under me. I dove and fired about 150 rounds; it burst into flames. This was in the region of Remonville at about 16 h 40. Confirmation requested – 1 Fokker, 1 – Balloon."

By the time hostilities ended on November 11, 1918, Rickenbacker was "America's Ace of Aces," being officially credited with 25 victories.[33] For his repeated acts of extraordinary heroism he was awarded the Medal of Honor; the Distinguished Service Cross with nine Oak Leaf clusters; the French Croix de Guerre with two palms; and the Chevalier de la Legion d'Honneur.

Rickenbacker was officially recommended for the Medal of Honor in 1919; however, it was not awarded to him until November 6, 1930, being presented by President Herbert Hoover, at Bolling Field, Washington, DC. Naturally some level of controversy surrounds this Medal because of the amount of time that elapsed. Some suggest the delay was a clear indication the U.S. military believed his actions did not warrant the award, adding by the late 1920s his popularity had grown considerably, reaching celebrity status, and that it had become politically correct to present it. However, others suggest the delay was simply the result of bureaucratic red tape. Whichever is correct, few could dispute that Rickenbacker's service during World War One was outstanding or that he had performed repeated deeds of extraordinary heroism that were above and beyond the call of duty.

His Medal of Honor Citation reads: "For conspicuous gallantry and intrepidity above and beyond the call of duty in action against the enemy near Billy, France, 25 September 1918. While on voluntary patrol over the lines, Lt. Rickenbacker attacked seven planes (Five Fokker type protecting two type Halberstadt). Disregarding the odds against him, he dived on them and shot down one of the Fokkers out of control. He then attacked one of the Halberstadts and sent it down also."

After the war, Rickenbacker returned to the United States. He wrote his wartime memoirs (*Fighting the Flying Circus*) and years later his autobiography (*Rickenbacker*). In 1922 he married Adelaide Durant and they had two sons. During WWII he actively toured various locations to meet

Captain Edward Rickenbacker standing in hangar doorway. The ribbon of his Distinguished Service Cross (left breast) displays five of his nine Oak Leaf clusters. Rickenbacker's DSC total remains the most ever awarded to one individual. (Photo George Williams.)

the fighting men. In 1942 the airplane he was flying ran out of gas and came down in the Pacific Ocean, 600 miles from Samoa. After three weeks on a raft, he and two other survivors were found and rescued. For his war service he received the Certificate of Merit. Rickenbacker remained in the public eye for the rest of his life, always being associated with cars and aviation until his death on July 27, 1973.

Robert Robinson & Ralph Talbot

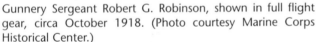

Gunnery Sergeant Robert G. Robinson, shown in full flight gear, circa October 1918. (Photo courtesy Marine Corps Historical Center.)

2nd Lieutenant Ralph Talbot, circa June 1918. (Photo courtesy Marine Corps Historical Center.)

The most aggressive aerial operation undertaken by the U.S. Marine Corps during World War One had as its primary mission the destruction of the German submarine bases along the Belgian coast – Bruges, Ostend, and Zeebrugge. This operation was the product of several Allied strategists, who suggested these enemy installations could be subjected to continuous day and night bombing by U.S. Marine and Navy aviation squadrons – collectively known as the Northern Bombing Group – based in the Calais-Dunkirk area. The plan sounded fine in theory, and the Marines and Navy embraced the operation enthusiastically, putting forth a great effort towards supplying the necessary personnel and equipment. But despite these efforts, the operation fell short in execution. Serious delays in aircraft deliveries greatly curtailed the group's ability to operate. In fact, Allied advances on the ground had brought the war to an end before this unit's air offensive really began. Yet in a sense, the Northern Bombing Group took on a symbolic importance that

outweighed its strategic impact. The unit took advantage of every resource available and created opportunities to engage enemy forces. In all, the group dropped nearly 78 tons of bombs on enemy targets before hostilities ended. And on several occasions, aviators forcefully engaged German aircraft in aerial combat. Two Marines in particular – Second Lieutenant Ralph Talbot (pilot) and Gunnery Sergeant Robert Robinson (observer) – were recognized for their 'exceptionally meritorious service and extraordinary heroism'; both being awarded the Medal of Honor for their actions.

Robert G. Robinson was born in Wayne, Michigan, on April 30, 1896. He was described as a soft spoken, likable young man, but quite determined. Robinson enlisted in the Marines on May 22, 1917, at Port Royal, South Carolina, and was subsequently transferred to the 92nd Marine Company at Quantico, Virginia. On June 17, 1918, he was attached to Squadron "C" of the First Marine Aviation Force, then located in Miami, Florida. Selected as an aerial gunner, Robinson was placed on detached duty and sent to the Armorer "gunnery" School at Fairfield, Ohio, on July 2, 1918. He completed the course on July 5, qualifying as an expert rifleman, and was promoted to corporal.

Ralph Talbot was born in South Weymouth, Massachusetts, on January 6, 1897. His contemporaries described him as a personable young man with keen gray eyes and light brown hair. He stature was typical – he weighed in at 148 pounds and stood nearly five-feet-ten-inches tall. Yet, many would add Talbot possessed an athletic look, having well defined muscles.

Talbot excelled both academically and athletically. While at Weymouth High School his sports record was outstanding and he showed rare talents in both public speaking and literature. "In fact, his ability along these lines was recognized by the faculty when they selected him to be orator for his class on the day of his graduation."[1]

Talbot entered Mercersburg Academy in Pennsylvania during the fall of 1915. But only for a short stay; "his academic ability, especially along literary lines, together with his athletic prowess, earned for him an opportunity to attend Yale University."[2] Arriving in the fall of 1916, Talbot participated as a member of the school's football, baseball, and cross-country teams in his freshman year.

When America entered the war in 1917, Talbot seems to have exemplified the image of the 'all American boy,' having joined Yale University's Artillery Training Corps. But in truth, his participation in a ROTC program was quite common during that time, and many patriotic men, eager for adventure, were quickly joining military schools throughout the country.

While in ROTC training Talbot was first exposed to aviation, but it quickly developed into a passion and he joined the Dupont Flying School, located in Wilmington, Delaware. In pursuit of his interest, Talbot enlisted in the Navy on October 26, 1917. He was appointed the rank of seamen second class and placed in the Navy's aviation program. His training began with ground school at the Massachusetts Institute of Technology (MIT). Talbot completed this course on January 1, 1918, and was promoted to chief-quartermaster. Then on April 8, he was appointed to the rank of Ensign and transferred to flight training at the Naval Air Station, Key West, Florida. Talbot successfully completed this course and was designated Naval Aviator No.456.

Coinciding with Talbot's training, both Marine and Navy aviation programs were expanding rapidly in preparation for war. And while the Navy had an abundance of aviators to meet expanding needs, the Marines had trouble filling their ranks until a number of Naval pilots realized the Marine program offered a greater opportunity of deployment overseas and transferred to the Corps. Adventurous by nature, Talbot was easily recruited into the Marine Corps, being honorably discharged from the Navy on May 17, and appointed Second Lieutenant (Marine Corps Reserve) the following day. Talbot arrived at the First Marine Aviation Force at Miami, Florida on May 26, and was assigned to active service with squadron "C"

Captain Alfred A. Cunningham standing in front of a Curtiss Jenny. Cunningham is justly considered the father of Marine Corps aviation for his initiative and drive in setting up training procedures and in selecting bases, equipment, and personnel for Corps aviation at its earliest stages.

as the Force mobilized for assignment in France.

It is unlikely Robinson and Talbot became acquainted while preparing for overseas deployment. Yet both men were members of the First Marine Aviation Force that sailed from New York aboard the U.S.S. De Kalb on July 18. The foundation for this force and Marine Corps Aviation in general is an interesting story which can be traced back to May 22, 1912, when First Lieutenant Alfred A. Cunningham reported for flight training at the Naval Aviation Camp near Annapolis, Maryland. In September 1915, he became Marine Aviator No.1 (Naval Aviator No.5), and a few weeks later he became the first Marine pilot to take off from a catapult at sea.[3] Cunningham is justly considered the "father of Marine Corps aviation" for his initiative and drive in setting up training procedures and in selecting bases, equipment, and personnel for Corps aviation at its earliest stages.

By the time the United States entered World War One on April 6, 1917, the Marine Corps had established an 'Aviation Section' at the Naval Aeronautic Station,

2nd Lieutenant Ralph Talbot in flying gear, standing in front of a Curtiss JN4-D training plane at the Marine Flying Field, Miami, Florida, June 1918. (Photo courtesy U.S. Marine Corps Historical Center.)

Pensacola, Florida – its rolls listed seven officers, 43 enlisted personnel, and four aircraft (Curtiss AH hydroplanes). Admittedly this unit was small, yet it would be wrong to blame any Marine. This was the product of a pre-war bureaucratic system that seemed to prioritize the Navy's Aviation program. As a general rule the Navy fliers were trained first and better equipped, while Marines waited at the end of the line.[4]

Nonetheless, within a few weeks the Corps authorized the organization of an Aeronautic Company under the command of Captain Cunningham and stationed this unit at the Marine Barracks, Philadelphia Navy Yard. Although personnel for this second unit were obtained through several sources, the primarily resource was the Marine Aviation Section at Pensacola, Florida.

Initially, the Marine's facilities at the Philadelphia Navy Yard consisted of only one hangar and three aircraft, two Curtiss R-6 seaplanes and one antiquated Farman land-plane.[5] But despite these humble beginnings, the Marine Aeronautic Company quickly initiated an aggressive flight training program and during the next six months the unit expanded to 34 officers and 330 enlisted men.

With sufficient forces now available, various assignments were entrusted to Corps aviation. To meet these demands, on October 14 the Marine Aeronautic Company was divided into two units; the First Aviation Squadron, consisting of 24 officers and 237 enlisted personnel; and the First Marine Aeronautic Company, consisting of 10 officers and 93 enlisted personnel. The following day, the First Marine Aeronautic Company was transferred to the Navy Air Station at Cape May, New Jersey; their main purpose was training in anti-submarine patrols. Despite the fact the unit had only two R-6 Curtiss seaplanes for this purpose, Marine training, with its discipline and drill, paid handsome dividends towards the Corps' image – the First Marine Aeronautic Company was to have the distinction of being the first American aviation unit of any service to be completely equipped and fully trained to leave the United States for service overseas.

The First Marine Aeronautic Company was detailed for duty in the Azores, with the primary mission of establishing a base from which they could fly anti-submarine patrols over the convoy routes. On January 9, 1918, the unit departed from Philadelphia aboard the USS *Hancock*, arriving at U.S. Naval Base No.13 in the Azores (Ponta Delgada, San Miguel Island) soon after. With them were 12 Curtiss Seaplanes – ten type R-6s and two type N-9s. Later, six HS-2L-6 flying boats were added to the unit's inventory.

Meanwhile, the First Aviation Squadron was transferred from the Philadelphia Navy Yard to the Army's airfield at Mineola, Long Island. However, their stay was short; inclement weather made the Marines realize the importance of establishing an airfield conducive to year-round flying. In pursuit of this goal, the squadron relocated to the warmer climate of Lake Charles, Louisiana, where flight training was resumed in early January 1918.

Coinciding with these activities, several American officers – including Cunningham – were sent on temporary assignment overseas to analyze various options for deployment in Belgium and Northern France. During the winter of 1918, most of these men returned to the States. Subsequently, several reports were submitted that outlined specific dilemmas the Allies were facing on these fronts. A number of these documents outlined one particular situation where enemy submarines operating from bases along the Belgium coast – Ostend, Zeebrugge, and Bruges – were causing great losses to Allied shipping. In part, one report strongly suggested that "...bombing raids upon U-Boat harbors will be not only necessary but the only way to meet and check the U-Boat menace."[6]

Admittedly, retaliatory action against these targets should have utilized established airgroups already operating within that sector, thereby affording the Allies a faster, more strategic response. Nevertheless, within a short time, both the U.S. Marines and Navy submitted proposals that suggested their respective aviation sections be allowed to set up bases and commence action against these important targets.

As these two branches of the U.S. military jockeyed for control of the operation, tension mounted. And perhaps in an effort to defuse the situation, a compromise was adopted on April 30, 1918, when the Navy Department

2nd Lieutenant Ralph Talbot, standing in front of a Curtiss JN4-D training plane at the Marine Flying Field, Miami, Florida, June 1918. (Photo courtesy U.S. Marine Corps Historical Center.)

authorized the development of a special Marine-Navy aviation force designated the Northern Bombing Group (NBG) and directed that all appropriate parties expedite assembly of the necessary personnel and equipment to support the effort.

The NBG was originally planned to operate with two wings – one day and one night – consisting of six squadrons each, and one assembly, repair, and supply unit, to be known as Base B, located in the close vicinity of these units. The day wing was to be composed of Marine squadrons, each equipped with 18 American DH-4 planes (Liberty Planes), and the night wing was to be composed of Navy squadrons, each equipped with ten British Handley Page bombers. However, soon after committing to this effort, the Navy came to the realization that the American DH-4 and British Handley Page programs were behind schedule. So, owing to the inability to obtain

sufficient numbers of aircraft, the Navy Department quickly scaled down the NBG to only four day and four night bomber squadrons.

With the reduction in force, however, the Navy now felt it had an abundance of personnel to support the NBG's night wing requirements. Therefore a policy was adopted which indicated qualified Navy pilots could transfer to the Marine's aviation program, which was still suffering from shortages in personnel. Fortunately, a sufficient number of Naval pilots believed a transfer to the Corps offered a greater opportunity of flying and deployment overseas. And soon the Marine's Day Wing squadrons were fully staffed.

As the Marines filled their ranks, training became the primary concern. To that end, Captain Geiger was selected to organize and train the Marines that would make up the NBG's day wing. In pursuit of this task, on February 7 the

The first American-built DH-4 arrived at La Fresne aerodrome from Pauillac, France, on 7 September of 1918. No organization insignia or unit markings had yet been added to this airplane. A wind generator is attached to the fuselage. (Photo courtesy National Archives.)

remaining Marines departed the Philadelphia Navy Yard for Miami, Florida. Three days later this group took over the Curtiss flying field located on the outskirts of Miami, quickly expanding it into a Marine Corps aviation training camp. Initially, the camp possessed only six JN-4 'Jennies.' However, Cunningham, who was then serving in Washington, D.C., as the first Officer-in-Charge of Marine Aviation, successfully lobbied for an additional 20 'Jennies' to be sent to Florida.

Conditions in Miami were less than favorable, yet the Marines had established a training facility that was large enough to accommodate all their aviation personnel in one location. Accordingly, on April 1, the First Aviation Squadron departed Louisiana for training in Florida. Once there, the entire group was reorganized into the First Marine Aviation Force with Cunningham in command. During the next two months the Force trained extensively as one organization. Then on June 16, the unit was formally broken down into a Headquarters Detachment and four Squadrons – designated A, B, C, and D.

On July 10, the Force was "detailed for foreign expeditionary shore service." And on July 18, a portion of the Force sailed on the USS De Kalb from New York, anchoring at Brest, France, 12 days later. The remaining portion, Squadron "D," arrived in France on October 5, bringing the unit's overseas strength to 149 officers and 842 enlisted men.

The NBG established four airfields behind the French and Belgian lines in the Calais-Dunkirk region. Navy squadrons that made up the Night Bombing Wing utilized St. Ingleverts (Squadrons 1 and 2) and Champagne (Squadrons 3 and 4). The Marine squadrons that

composed the Day Bombing Wing were housed at Oye (Squadrons A and B) and La Fresne (Squadrons C and D).[7] Group headquarters was at Antingues, a few miles south of Ardres. Antingues also housed the field supply depot and motor transport park. A further site, at Bois-en-Ardres, was selected as day wing headquarters.

Although none of the NBG's squadrons possessed operational aircraft upon reaching their designated airfields, it would seem the Marines, in particular, were unable to resolve this issue. Of the 72 aircraft shipped to France for their use, only four of these planes – American manufactured DH-4s with Liberty motors – were assembled and available for use in early September. And as weeks passed, non-delivery of remaining DH-4s forced the NBG to obtain, by concession of the British Government, 54 DH-9A aircraft in exchange for American-built Liberty motors.[8]

A more successful story that coincided with efforts to obtain aircraft was the training program the Marines established to prepare their flying personnel. Primary training in flying, aerial gunnery, and bombing was provided at aviation schools at England and France, while first-hand experience was accomplished by placing Marine pilots and observers temporarily in active British squadrons at the front.

On August 9, 1918, the first three Marine crews were transferred temporarily to 218 Squadron, RAF, whose aerodrome was situated near Calais. Owing to the programs success, the number of Marine crews sent to the RAF pilots pool, at Audembert, France, was expanded in early September. After successful qualification, they were in turn transferred as needed to either 217 or 218

DH-9A obtained from the Royal Air Force and assigned to Squadron "C" of the NBG These aircraft were powered by Packard-built or Lincoln-built Liberty engines. (Photo courtesy National Archives.)

Squadron, RAF, and from there, after actual bombing experience, were sent back to the squadrons of the Marine day wing.

Talbot and Robinson were sent to the replacement pool at Audembert on September 15. Within a short time, Talbot qualified as pilot and Robinson as aerial gunner; soon after both men were transferred to 218 Squadron, RAF, in late September 1918.

This squadron was called into special service on the evening of October 1, after Allied commanders received word that several Belgian and French infantry divisions operating near Stadenburg had exhausted their food supplies. Due to unusable roads near their positions, re-supplying these units through ground transportation was impossible. In response to the situation, Belgian aviation headquarters was ordered to air-drop supplies. But, as the Belgian Air Force was short of aircraft, Royal Air Force units were placed at their disposal, including 218 Squadron.

The mission required 15,000 ration canisters, having a total weight of 13 tons, to be placed into hundreds of small bags that held five to ten ration canisters each. After considering weight and space issues, the aircrews loaded a quantity of these bags into the observer's cockpit of each participating plane. The observer threw the bags overboard once the pilot signaled the specified location.[9]

To ensure the supplies were accurately delivered, many of these aircrews drops their loads from an altitude of only 500 feet and therefore were subjected to intensive enemy ground fire. The Marines exemplified this on October 2,

when one crew – Captain Lytle and Gunnery-Sergeant Wiman – made ten low-level runs in the face of intense enemy fire while dropping 4,300 pounds of badly needed food and supplies to an isolated French regiment near Stadenburg. During several passes their DH-4 (designated Marine Aircraft No. D-1) was hit numerous times by enemy gun fire.

Talbot and Robinson made similar runs in aircraft D-1 on October 3, having relieved Lytle and Wiman, who at this point were in desperate need of rest. In later discussions, Talbot would lightly refer to these mission as his "aerial grocery business,"[10] but one of his comrades better described the dangers he encountered: "The tins of bully beef were dropped from under 500 feet, and the ships were exposed to a terrific machine gun and rifle fire."[11]

Marine Aircraft No. D-1 was one of the few American-built machines provided to the Marine's Day Wing. Powered by a Liberty motor, it had a top speed of 124 mph, making it the fastest aircraft in 218 Squadron, which was otherwise equipped with slower DH-9s. As a result, while Talbot and Robinson were obtaining their actual bombing experience with this unit, the two marines were generally relegated the dangerous task of protecting the formation flights from enemy attack.

In this role Talbot and Robinson participated on bombing raids over Belgium – Thourout on October 4; Oroye on October 5; and the railway yards at Westende, Nieuport, and Ostend on October 7. But in each case no enemy planes were encountered.

DH-4s and DH-9As at La Fresne aerodrome, circa fall 1918. (Photo courtesy U.S. Marine Corps Historical Center.)

Marine DH-4s at La Fresne aerodrome being readied for an early morning bombing mission, circa fall 1918 (Photo courtesy U.S. Marine Corps Historical Center.)

Luck seemed the major factor, because the aviators knew large numbers of German aircraft were present in their area of operation. Yet Talbot expressed concern over not seeing any enemy planes. Ironically, he had little time to contemplate these thoughts. The following day, October 8, while Talbot and Robinson participated in a bombing raid over Ardoye, Belgium, their flight was attacked by nine German fighters. As planned, the two marines were flying their DH-4 in a position to protect the Allied formation. However, the two men carried this role beyond what was expected when they quickly engaged the enemy machines single-handedly, despite the odds being nine to one.

Nevertheless, Talbot and Robinson were so aggressive in their attack they shot down one enemy aircraft before making good their own escape. Four days later Talbot

received a letter of commendation from Major Alfred A. Cunnungham, Day Wing Commander, which stated:

Bois en Ardres
12 October 1918

Subject Commendation.

1. Official confirmation has been received of the air combat in which you engaged on 8 October, 1918, and in which you shot down a German airplane. The Commanding Officer wishes to express his appreciation of your good work, of which the whole wing is justly proud, and assure you that he has no doubt but that you will further distinguish yourself.

On October 12, Talbot and Robinson officially ended

1/Lt. Herman A. Peterson (pilot) and 2/Lt. Charles R. Needham (observer), Marine personnel of Squadron "C," ready for take-off for a test flight of a DH-4. The Marine Corps insignia was usually added to the fuselage prior to the first combat flight. (Photo courtesy National Archives.)

their temporary assignment with the RAF by flying aircraft D-1 to Marine Squadron "C" at La Fresne aerodrome. In their absence, the Day Wing had finally obtained enough machines to commence independent bombing missions. In conjunction with this issue, two days later, on October 14, seven machines from Squadron "C" carried out the first all-Marine air combat operation for the NBG's Day Wing.

This flight was to attack the German-held railway center at Thielt, Belgium.[12] Although eight planes started out, due to engine trouble one was forced to turn back before reaching the front lines. The seven remaining aircraft flew in a close "V" formation, with the flight leader, Captain Lytle, positioned at the apex of the "V."

The Marine's attacked the designated target without incident – dropping over 2,000 pounds of bombs – but unfortunately, on their return flight the formation was intercepted by a group of 11 German fighters.[13] As these aircraft approached the Marine bombers head on, they split into two groups – four enemy planes turned on the right side of the Marine's "V" formation, attempting to attack the lead bomber; while the other enemy aircraft swept up on the left side of the "V" in an attempt to concentrate on those Marine planes.

Luckily, Lytle sized up the situation quickly and signaled his pilots to tighten up their battle formation. Meanwhile, the observers brought their guns to bear on the approaching enemy planes, opening fire as soon as they came within range. Lytle directed his observer, Wiman, to concentrate on the four enemy aircraft approaching on their right flank. At a range of 400 yards, Wiman fired about 25 rounds into the leading enemy fighter, causing its pilot to change his flight-path immediately, as he passed under the DH-4 at a range of 50 yards. While the first German fighter dove out of range, a second German attacked the Lytle-Wiman DH-4 from a position under its tail. At a range of 200 yards this enemy's gun fire hammered the wings and center section of the DH-4. The Marines responded swiftly; Lytle banked the DH-4, bringing this German aircraft into Wiman's field of fire, who in turn discharged a full drum of ammunition into it, causing the plane to go down apparently out of control.

At this point in the battle, two Marine aircraft began to detach from the "V" formation due to engine trouble. However, it seems only one of these troubled machines was selected as a target by the Germans – the DH-4 crewed by Talbot and Robinson – for a large number of enemy fighters converged on it. In defense, Robinson squeezed off several bursts from his machine gun, hitting one enemy aircraft which went down out of control. However, two other German fighters had moved into a position beneath the tail of the DH-4 and quickly attacked, firing up through the flooring. In the first few moments of this encounter a German bullet found its mark, shattering Robinson's left elbow so severely his arm hung by a single tendon.

DH-4 of Marine Squadron "C," No. D-13 (serial A-3279). Captain Robert S. Lytle (front cockpit) and Sergeant Amil Wyman (observer-gunner cockpit). This aircraft is armed with a single 0.30-caliber Lewis machine gun mounted in the observer's position and two forward-firing 0.30-caliber Marlin machine guns. Bomb racks can be seen under the lower starboard wing and the bomb sight is located near the observer's position.

Below: DH-4s of Marine Squadron "C" lined up at La Fresne aerodrome, circa fall 1918.

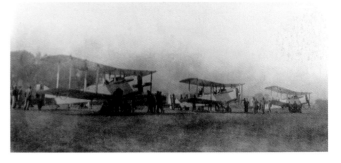

Despite the intense pain Robinson endured as his left arm dangled helpless at his side, the ferocity of the battle forced him to continue firing with only his right hand. And for a short time he succeeded; however, a cartridge soon jammed his gun. Had Robinson not been wounded, clearing a jammed gun under these battle conditions would have been very difficult. Considering his wound, the situation seemed grim. Fortunately, Talbot realized Robinson's plight, and in a defensive effort, quickly maneuvered the DH-4 around and headed towards an enemy aircraft in hopes of using his forward guns. But as he closed-in on this target while firing, his guns also jammed.

Talbot's maneuver did, however, afford the wounded Robinson the necessary time to clear the jammed cartridge from his machine gun. And he bravely returned to the fight, perhaps hitting a second enemy aircraft before he was shot twice more – once in the stomach and once in the thigh – causing him to collapse into the observer's cockpit, mercifully unconscious.

Unfortunately for the two marines, Robinson landed on top of the plane's control cables, making the DH-4 nearly impossible for Talbot to control, who at this point was attempting to turn the machine. Yet, with great difficulty, Talbot finished his turn. And then, despite the fact his front gun was still jammed, he flew directly at an enemy aircraft in such an aggressive manner it forced the German to retreat. This daring maneuver allowed Talbot an opportunity for escape, which he took full advantage of; diving his DH-4 towards the ground he eluded the balance of the enemy fighters while crossing the German trenches at an altitude of only 50 feet. Talbot was now over Allied lines, but his aircraft had sustained considerable combat damage and his observer was seriously wounded. The latter was paramount is Talbot's mind, and he continued

his flight to the Belgian aerodrome at Hondschoote, where a field hospital was located. There, the wounded Robinson was rushed to the field hospital. Soon his arm was successfully grafted back on by the surgeon-general of the Belgian Army.

Meanwhile, Talbot returned to his aerodrome alone. According to fellow aviator Second Lieutenant Alvin L. Prichard, USMC, the two men had a general discussion about the mission of October 14, which Prichard in turn summarized in a letter to Talbot's mother on October 26, 1918:

"…It developed that 11 Huns [Germans] had dived on him, that eight were too slow and remained firing from a distance, while three kept diving on him and firing. Robinson shot one down and his gun jammed. At that instant he received a stream of bullets, through his left arm, rendering it useless.

"While he was working the gun, Dick turned on the Huns [Germans] with his forward guns but after three or four shots, they too jammed. Then Robinson, recovered, shot down a second. The third dove and shot him through the chest and stomach.

"Dick turned on the third with all guns useless, one of the most daring, magnificent bluffs of the war. Then, with

Ground crew working on Marine aircraft No. D-9, serial No.32360 (Navy serial No.A-3264). (Photo courtesy National Archives.)

DH-4s of the First Marine Aviation Force lined up in France, late fall 1918. (Photo courtesy National Archives.)

Robinson having fainted and fallen on the controls, Dick dove to within a hundred feet of the ground and started back across miles of hostile territory, with the third Hun [German] on his tail, firing constantly. Dick's account to me was most graphic.

"Twisting, turning, zooming trees, he fled with every ounce of power, while above the roar of the motor he could feel the zip of the bullets as every part of the plane was struck.

"At every house Boches [Germans] would snipe at him

Painting depicting DH-9As of the Northern Bombing Group on a mission over enemy territory. (Photo courtesy Marine Corps Historical Center.)

with rifles, and as he passed under a barrage the Huns [Germans] were laying on the trenches, saw familiar uniforms, and the Hun scout left him.

"Again he showed his coolness by landing at a Belgian airdrome and rushed the unconscious Robinson to a hospital, undoubtedly saving his life. The Belgian major was very enthusiastic about his conduct, and it is understood, was going to recommend Dick and Robinson for medals for heroic conduct."[14]

Within days Talbot's aircraft was patched up and ready for action. So was he. And as the NBG's Day Wing bombed German railway centers, canals, supply dumps, and aerodromes, Talbot participated.

On October 17, two raids were carried out by the Marines. In the morning the railway yards at Steenbrugge were bombed by a formation of five aircraft from Squadron "C." A further raid was directed on Zeebrugge Mole and the Harbor at Ostend that afternoon. The following day an impressive raid was made by a formation of seven Marine bombers; their primary objective was the railway center at Eceloo, Belgium. Despite heavy anti-aircraft fire, the aircraft were brought down to a height of 3,000 feet while dropping their bombs on the railway yards. Several direct hits were observed. It was later ascertained that one was made on a troop train, resulting in the death of 60 enemy officers and 300 enlisted men.

Next the marines sighted a German aerodrome about three miles southeast of Eceloo. Noting 12 enemy aircraft on the field – presumably Gotha bombers – the marines dropped their remaining bombs on this aerodrome.

Further air raids by the Marines followed in late October 1918; however, these missions were greatly hampered by a dwindling supply of serviceable aircraft. To make matters worse, the continuous flying required throughout the month had clearly pushed many of the machines beyond their normal life span, but no replacements were available. This issue was best illustrated on October 25, when Talbot (pilot) and Second Lieutenant Colgate W. Darden Jr., (observer) attempted an engine test flight in worn-out DH-4 (Marine Aircraft Number D-1) on airfield E, at North Bowling, France.

On the first run, Talbot raced the plane down the airfield, but it failed to generate enough power for takeoff. During the second run, Talbot twice attempted liftoff but rebounded on the ground as it failed to gain altitude. He was about three feet above the ground when he failed to clear a bomb embankment trench at the end of the field. The aircraft's landing gear was ripped off on impact, causing the plane to flip over and crash into a large pile of bombs.

Darden was thrown clear of the plane as it flipped over. Although seriously injured, he survived. Talbot was not as fortunate; he was trapped in the cockpit as the plane exploded in flames.

An eyewitness to the events of October 26, was George C. Connor, USMC, who stated: "Talbot took off about 9 or 10 o'clock in the morning with a man in the rear cockpit and had taxied by our camp, which was in the apple orchard, half way between Calais and Dunkirk.

"He had just cleared the ground and was straightening out. I happened to be standing on the outskirts of our camp, when I heard his motor cease roaring and hesitate in the air for a second. Then I saw it make a nose dive down into a bomb trench our men had been digging the day before.

"The man in the rear cockpit was seen to shoot up out of the plane and come down among the racks of bombs, while a great burst of flames showed that the gas tanks had broken away and the plane was a mass of flames 40 feet high in an instant.

Dramatic crash scene at La Fresne aerodrome on 25 October 1918, in which 2nd Lt. Ralph Talbot was killed and Marine aircraft No. D-1 was destroyed. Liberty engine and structure of upper right wing can be seen in this photo. Rescue workers are seen trying to put out intense flames. (Photo courtesy of Marine Corps Historical Center.)

"The first sergeant was already dashing out over the field, a distance of about a thousand feet from our hidden tent camp in the orchard.

"When I came up, the first sergeant was already down in the bomb trench trying to push the plane around so the flames would be kept off Talbot. The pans of Lewis machine gun ammunition were being sprayed in all directions by the flames, although the ammunition was bursting rather than shooting.

"Men began running from the camp and one sergeant had the presence of mind to dash into a hangar and come running out in the little flivver ambulance. Half a dozen fire extinguishers were in use, some on Talbot's plane but mostly on the racks of bombs, which in their wooden castings with detonators attached were beginning to burn fiercely.

"While some of the men put out the flames, others grabbed the bombs in their arms and carried them 30 or 40 feet away, piled them on the ground. Some of those bombs were smoldering in our arms as we carried them away.

"By the time we got the fire out, or at least were able to get to Ralph, the poor boy could be seen with the heavy engine lying on his chest and his hands and feet practically burned off. I think that Ralph died instantly and was not tortured by the fire. The big engine was squarely on his chest and must have killed him when the plane crashed in less than a second.

"The running gear of the plane had caught on the embankment of earth thrown out of the bomb trench. It crashed at not less than six feet off the ground.

"He used to come into my quarters at night, when we were all darkened down to keep the Heinies [Germans] from dropping their pills [bombs] on us, and tell me about the towns behind the lines.

"He said that when he landed in Ghent, Belgium, just ahead of the British flyers, the inhabitants rushed out and grabbed his arms and his legs and kissed his uniform, cried and danced and went on like crazy people.

"Ralph Talbot – I will ever remember his kindly gray eyes and the crinkles about them when he laughed, and he was generally doing that. He had two fistfuls of busted fingertips from playing baseball in his school days and he was some player too. We often played a game of baseball to relieve the tension of strained nerves."[15]

On 29 October 1918, Ralph Talbot was buried with full military honors in the British Military Cemetery at Les Baracques near Calais – Pilot No.5, Row "B," Grave No.3. His body was later moved to a cemetery in his home town of South Weymouth, Massachusetts.

Talbot was awarded the Medal of Honor, posthumously, on November 11, 1920. His citation reads: "For exceptionally meritorious service and extraordinary heroism while attached to Squadron C, First Marine Aviation Force, in France. Second Lieutenant Talbot participated in numerous air raids into enemy territory. On 8 October 1918, while on such a raid, he was attacked by nine enemy scouts, and in the fight that followed shot down an enemy plane. Also, on 14 October 1918, while on a raid over Pittham, Belgium, Lieutenant Talbot and another plane became detached from the formation on account of motor trouble, and were attacked by 12 enemy scouts. During the severe fight that followed, his plane shot down one of the enemy scouts. His observer was shot through the elbow and his gun jammed. Second Lieutenant Talbot maneuvered to gain time for his observer to clear the jam with one hand, and then returned to the fight. The observer fought until shot twice, once in the stomach and once in the hip. When he collapsed, Lieutenant Talbot attacked the nearest enemy scout with his front guns and shot him down. With his observer unconscious and his motor failing, he dived to escape the balance of the enemy and crossed the German trenches at an altitude of 50 feet, landing at the nearest hospital to leave his observer, and then returning to his aerodrome."

In October 1936, a Navy destroyer – USS *Ralph Talbot,* was commissioned in his honor at the Charlestown Navy Yard.

At the time of Talbot's death, Robinson was still recovering from his wounds. He was transferred to U.S.

Wreckage of the plane in which Talbot met his death. The photograph was taken shortly after the fatal crash at La Fresne aerodrome. Clearly visible is the dirt embankment into which the plane crashed. Bombs for the ammunition dump have been removed and rescue workers have extinguished the flames. Undercarriage, engine, and wings are still visible. (Photo courtesy of Marine Corps Historical Center.)

Naval Hospital at Brest, France, on November 28, 1918, and to the Marine Barracks, New York, on January 6, 1919. Robinson was honorably discharged on June 16 as Gunnery Sergeant and appointed a Second Lieutenant, class 5, in the Marine Corps Reserve at Washington, D.C., on June 17, being assigned to inactive duty. His retirement was effected in May 1923, and his promotion to First Lieutenant occurred in September 1936. Robinson was also awarded the Medal of Honor for his efforts during October 1918. His citation, dated November 11, 1920, reads the same as Talbot's.

Upon retirement, Robinson made his home at St. Ignace, Michigan. He lived to see the initiation of the Robert G. Robinson Award in 1971, being presented annually to the "Marine Flight Officer of the Year." The award symbolized the quiet strength and determination of the Gunnery Sergeant who earned the Medal of Honor against severe odds. He died at his home on October 5, 1974, and was buried with full military honors in Arlington national Cemetery, Arlington, Virginia.

The Destroyer U.S.S. *Ralph Talbot* was commissioned in October 1936. (Photo courtesy U.S. Marine Corps Historical Center.)

Endnotes

Introduction
1. Between 1866 and 1940.
2. The Congressional act which created the Army's Medal on July 12, 1862, 1861, specified gallantry in action "and other soldierlike qualities" as the basis for the award.
3. *The Congressional Medal of Honor – The Names, The Deeds,* Sharp & Dunnigan, California, 1984. p.3.
4. The cost of the medal was two dollars each. *Above and Beyond,* Boston Publishing Company, Massachusetts, 1985. p.5.
5. *The Congressional Medal of Honor — The Names, The Deeds,* Sharp & Dunnigan, California, 1984, p.7.

Bleckley and Goettler
1. *Cross & Cockade Journal,* Vol. 7, p.365.
2. Daniel P. Morse, Jr., *The History of the 50th Aero Squadron – The "Dutch Girl Observation Squadron in World War I,* The Battery Press, Tennessee, 1990, p.23. "each pilot being assigned his own machine."
3. During this time members of the 50th Aero were promoted in rank, Bleckley and Goettler to first lieutenant on September 17, for their excellent work during the St. Mihiel Offensive.
4. Although the operation ended on September 26, 1918, during the next three days aerial patrols were generally dispatched only for the purpose of keeping up a light surveillance of the enemy sector.
4. *Cross & Cockade Journal,* Vol. 8, p.85.
5. It is not known which two aircraft from the 50th Aero Squadron carried the 77th Division's "Statue of Liberty" insignia. It is assumed the markings were crude because they were applied to the aircraft after the 50th Aero Squadron moved to Remicourt on September 24, 1918. This was only two days before the offensive began.
6. *Cher Ami* (Dear Friend) was the battalion's last pigeon. His flight from the Charlevaux Valley on October 5, 1918, to the 77th Division's pigeon loft at Rampont took about fifteen minutes. The bird's leg had been shattered, one wing badly injured, and its breastbone was broken by shell fire, but the message was there, dangling from the capsule on his leg.
7. The crews and aircraft used that day were: Goettler/Bleckley, Number "2" and "6" planes; Pickrell/George, "Number 6" plane; Bird/Bolt, "Number 14" plane; Phillips/Brown, "Number 8" plane; McCurdy/Graham, "Number 1" plane; Anderson/Roger, "Number 16" plane; Beebe, "Number 4" plane; Slater, "Number 12" plane; Frayne, "Number 5" plane. Other participants included McCook, Lockwood, Brill, Batson, Thomson, Sain, Dovre, and the French. However, the particular aircraft numbers used by them are not known.
8. Of the 600 men who entered the ravine on October 2, only 194 marched out on October 8. 170 were dead; the rest were carried out on stretchers, with many dying of their wounds in army hospitals. Whittlesey and three of his men received the Medal of Honor for their actions between October 2–6, 1918.

Hammann
1. Francis Ormsbee was the other U.S. Naval aviator who received the Medal of Honor during World War One. He earned it while stationed stateside at NAS Pensacola, Florida, for an act of bravery not involving conflict with enemy forces.
2. Dennis Gordon & Martin O'Connor, "Worried Landings," *Naval History,* April 1987, p. 24. The squadron's surgeon and two enlisted gunners were also awarded the Navy Cross.
3. Landsman for Quartermaster was a rate given to an enlisted man who served on a warship without any prior naval training. Typically, these men were volunteers whose naval service ended with the end of hostilities.
4. Dennis Gordon & Martin O'Connor, "Worried Landings," *Naval History,* April 1987, p. 24. First Lieutenant, John L Callan, USNRF, recommendations on the air stations [San Severo, Pescara, and Porto Corsini] and school [Bolsena] were set to paper on 26 January, 1918, in a letter from him to the Commander, U.S. Naval Aviation Forces in Foreign Service.
5. Lieutenant Commander H. H. Frost, U.S.N., "An Air Adventure Over Pola," *Sea Power,* April 1921, p. 180.
6. Noel Shirley, "John Lansing Callan – Naval Aviation Pioneer," *Over The Front,* vol.2, no.2, 1987, p.183.
7. It was initially envisioned that for Porto Corsini there would be 18 F.B.A. type H flying boats for patrolling and bombing, six Ansaldo S.V.A. 5 floatplanes for defending the station against air attacks, and 12 Macchi-built single-seat flying boats for escorting the patrol machines and for quick reconnaissance work.
8. James J Sloan, Jr., *Wings of Honor,* PA, 1994, p.263.
9. Lieutenant Commander H. H. Frost, U.S.N., "An Air Adventure Over Pola," *Sea Power,* April 1921, p.181.
10. Dennis Gordon & Martin O'Connor, "Worried Landings," *Naval History,* April 1987, p.24.
11. Dennis Gordon & Martin O'Connor, "Worried Landings," *Naval History,* April 1987, p.26.

Luke
1. Harold Hartney, *Up & At'Em,* New York, 1940, p.209
2. Ibid., p.209
3. Ibid., p.208
4. While at San Diego, California, Luke met and became engaged to Miss Marie Rapson.
5. The First Pursuit Group was compromised of the 27th, 94th, 95th, and 147th Aero Squadrons.
6. Although the 27th Aero claimed four victories, *JG.1* recorded no losses this day. In return the Germans claimed four Nieuports – one to *Ltn.* Ernst Udet of *Jasta* 4, two to *Ltn.* Erich Lowenhardt of *Jasta* 10, and one to *Ltn.* Fritz Friedrichs of *Jasta* 10. Yet, two of the Americans did in fact make it back to their lines safely.
7. Udet's victim – 2nd Lt. Walter B. Wanamaker – was brought down in Nieuport 28 N6347 (No.3) and taken prisoner.
8. Among the losses were 1st Lts. Charles B Sands and Jason S. Hunt – both killed; while 1st Lt. R.C. Martin was taken prisoner. 1st Lt. Oliver T. Beauchamp made it back to the aerodrome, only to fatally crash while

landing. The fifth loss was 1st Lt. Charles A. McElvain, who became separated from his flight and engaged a group of Fokkers over Soissons. Although he shot down the flight leader – *Leutnant der Reserve* Gunther Schuster, a six-victory ace and commander of *Jasta* 17, McElvain was himself shot down near Arcy by another German – *Ltn. d.R.* Alfred Fleischer – and made prisoner.

9. Although Hartney claims to have lead this patrol, squadron reports do support this claim.

10. Harold Hartney, *Up & At'Em,* New York, 1940, p.214

11. Ibid., p.214

12. Ibid., p.214

13. Wehner had finally achieved a take-off, but was unable to locate the formation or Luke. Nevertheless, he did find and destroy an enemy balloon south-west of Spincourt before landing.

14. Edward Rickenbacker, *Fighting The Flying Circus,* New York, 1965, p. 196

15. Ibid., p.196

16. Ibid., p.196

17. Ibid., p.196

18. Harold Hartney, *Up & At'Em,* New York, 1940, p.225

19. Ibid., p.225

20. Jon Guttman, "Balloon Buster", *Aviation History,* p. 46

21. Jon Guttman, "Balloon Buster", *Aviation History,* p. 51

22. Jon Guttman, "France's Foreign Legion of the Air, No.3: Reginal Sinclair and the Hunting Horns of SPA 68," *Windsock International* (UK) vol.5 no.3, 1989, p.23

23. Harold Hartney, *Up & At'Em,* New York, 1940, p.231

24. Ibid., p.231

25. The 94th Aero Squadron would end the war as the top scoring squadron of the AEF.

26. Luke's first Distinguished Service Cross was recommended by Hartney on September 17, 1918. Quoting from his letter: "The work in which this officer had displayed such extreme heroism is of particular military value that tends to blind the enemy air service at the most important time during the enemy's evacuation. This officer, on almost every occasion, has had his plane damaged by machine gun fire but undaunted, searched out, locates balloon and pressed home these successive attacks without fear and with the utmost skill and gallantry. He has volunteered on many occasions to undertake this work when anyone would have excused him for not flying."

27. Jon Guttman, "Balloon Buster", Aviation History, p. 51

28. James Parks, "Jerry Cox Vasconcells, Colorado's Only A.E.F. Ace", *Over The Front,* vol.3 no.4, 1988, pp. 339–340. Lt. Russell Pruden, the Supply Officer for the 27th, was at the advanced aerodrome near Verdun. He writes in his diary for September 29: "Rainy and misty all day. No patrol. Maj. Hartney and Lt. Nicholson, one of the new pilots landed here during the day. Luke landed early in the afternoon with engine trouble and phoned down to the main field that he would stay and go up after balloons about 1700. He also sent word over to our balloons to watch for him and get his confirmations."

29. Harold Hartney, *Up & At'Em,* New York, 1940, p.235.

Hartney admits "Luke had not asked me, but knowing his mind and his ability, I said: Yes..."

30. Ibid., p.235

31. Ibid., p.340

32. James Parks, "Jerry Cox Vasconcells, Colorado's Only A.E.F. Ace", *Over The Front,* vol.3 no.4, 1988, pp.340. Lt. Russell Pruden, writes in his diary for September 29: "He [Luke] started at 1700 but landed again with more motor trouble. Started a second time at about 1900."

33. Jon Guttman, "Balloon Buster", *Aviation History,* p. 55

Ormsbee

1. When the Secretary of the Navy approved the Order on December 3, 1918, Ormsbee was officially acknowledged as a recipient of the Medal of Honor. However, the award was not pinned on Ormsbee until later ceremonies.

Some argue that the basis for determining the first aviator awarded the Medal of Honor should be the time and date on which the individual performed the specific act of bravery. By those conditions, Charles Hammann's actions on August 21, 1918, would make him the first aviator awarded the Medal of Honor. Edward Rickenbacker's actions of 8:40a.m., September 25, 1918, would make him the second aviator awarded the Medal. Ormsbee's actions of 10:10a.m., September 25, 1918, occurred hours after Rickenbacker (allowing for the time zone difference), and make him third on the list.

Rickenbacker

1. Rickenbacker was not officially credited with the victory of May 7, 1918, until 1960.

2. Edward Rickenbacker, *Rickenbacker An Autobiography,* Prentice-Hall Inc., New Jersey, 1967, p.21.

3. Ibid., p.85.

4. Ibid., p.88.

5. Ibid., p.90.

6. Ibid., p.91.

7. Ibid., p.91.

8. Jack Eder, *Let's Go Where The Action Is! – The Wartime Experiences Of Douglas Campbell,* JaaRE Publishing, Indiana, 1984, pp.41–43.

9. Major Jean W.F.M. Huffer, a flyer of considerable reputation, had served with the Lafayette Escadrille.

10. Jack Eder, *Let's Go Where The Action Is! – The Wartime Experiences Of Douglas Campbell,* JaaRE Publishing, Indiana, 1984, p.50.

11. Edward Rickenbacker, *Rickenbacker An Autobiography,* Prentice-Hall Inc., New Jersey, 1967, p.99. Anti-aircraft fire was known as 'ack' ack" or 'flak," it was universally known as "Archie" in World War One, from a popular English song that ended with the words: "Archibald! Certainly Not!"

12. Edward Rickenbacker, *Fighting The Flying Circus,* New York, 1965, p.5.

13. Ibid, p.7.

14. Ibid., p.7.

15. Edward Rickenbacker, *Rickenbacker An Autobiography,* Prentice-Hall Inc., New Jersey, 1967, p.100.

16. Jack Eder, *Let's Go Where The Action Is! – The Wartime Experiences Of Douglas Campbell,* JaaRE Publishing,

Indiana, 1984, p.51.

17. When Hall was finally released in 1918, he reported Scheerer's death, and credit for bringing him down was belatedly awarded to Rickenbacker in 1960! After the war Hall co-authored with Charles Nordhoff *The Lafayette Flying Corps,* the official history of the unit. He had numerous best sellers over the years, including the classic *Mutiny on the Bounty.*

18. A vrille was a "tail spin," or simply a spin, used by pilots as a ruse to make opposing pilots think the plane was going down out of control. The pilot would then recover from the spin at a lower altitude and escape.

19. Edward Rickenbacker, *Fighting The Flying Circus,* New York, 1965, p.61.

20. Ibid., p.61.

21. Edward Rickenbacker, *Rickenbacker An Autobiography,* Prentice-Hall Inc., New Jersey, 1967, p.110.

22. Edward Rickenbacker, *Rickenbacker An Autobiography,* Prentice-Hall Inc., New Jersey, 1967, p.111.

23. Rickenbacker, Green, and Loomis claimed a German two-seater on July 8, 1918. However, the victory was never confirmed.

24. Rickenbacker claimed a German fighter on August 10, stating: "I got on the tail of a Boche [German] and got two bursts of 75 shots into him he fell off into a spiral. I could not watch him because I was kept busy by other enemy planes. I received a ball in one of my wings." Rickenbacker's SPAD 13 received a bullet patch, but the victory was never confirmed.

25. Edward Rickenbacker, *Fighting The Flying Circus,* New York, 1965, p.207.

26. A reversement is now known as an Immelmann, named after the German ace who invented it. It is a half loop followed by a half roll; the aircraft reverses direction and gains altitude.

27. Ibid., p208.

28. Ibid., p208. Apparently in later years, Rickenbacker mistakenly described the photographic machine as a L.V.G. and that it had been shot down in flames. His September 25, 1918 combat report and Medal of Honor citation both state the two-seater was a Halberstadt. Furthermore, neither mentions the two-seater going down in flames.

29. Edward Rickenbacker, *Rickenbacker An Autobiography,* Prentice-Hall Inc., New Jersey, 1967, p.128.

30. Ibid., p.128.

31. This two-seater was incorrectly listed as a Halberstadt. However, being captured intact, it was later documented as Hannover CL.IIIa serial 3892/18.

32. *Cross & Cockade Journal,* Vol. 3, No. 3, p.271.

33. Rickenbacker was not officially credited with the victory of May 7, 1918, until 1960.

Robinson & Talbot

1. Memorial Program of the Launching of the U.S.S. *Ralph Talbot,* 1936, p.5

2. Ibid., p.5–6.

3. His Aero Club of America Seaplane Certificate was No.2, received at the Burgess Aviation School in 1913.

4. Ironically, such favoritism plagued Marine aviation during the earliest days of World War Two.

5. For the longest time, the only additional aircraft were two Curtiss JN-4 'Jennie' biplanes.

6. James J. Sloan, Jr., *Wings Of Honor,* Schiffer Military/ Aviation History, PA, 1994, p.266.

7. The Marine squadrons that made up the Day Wing were sometimes referred to a squadrons 9, 10, 11, and 12.

8. Despite British help the situation never really improved. By the signing of the Armistice on November 11, aircraft on hand for the Marines were only 12 DH-4s (8 in commission), and 17 DH-9s (7 in commission), considerably under the planned number of 72 aircraft.

9. In total, 80 Allied planes participated in this relief mission.

10. Memorial Program of the Launching of the U.S.S. *Ralph Talbot,* 1936, p.7

11. Ibid., p.6–7.

12. Some sources referred to the objective of this bombing raid as being the "railroad junction at Thielt Rivy, Belguim" rather than "an air raid over Pittham, Belguim" as stated in the citation. It is just possible that "Rivy" was meant to be "Rwy" as an abbreviation for railway, and that the two descriptions refer to the same objective.

13. Several sources suggest a mix of Fokker and Pfalz fighters, but the aircraft types are not known for certain.

14. Memorial Program of the Launching of the U.S.S. *Ralph Talbot,* 1936, p.7–8. Prichard (Naval Aviator No.279) was a pilot with Squadron "C" and Talbot's tentmate from the day they landed in France.

15. *Philadelphia Bulletin,* February 8, 1928.

Color Section

1. Rickenbacker, Edward V.; *Fighting the Flying Circus;* Frederick A. Stocks Company, New York, 1919. p.16.

2. Jon Guttman, Allan D. Toelle, Howard Fisher, Greg Van Wyngarden; *94th Aero Squadron "Hat in the Ring"* Part 1: (Nieuports at Toul); Over the Front; Vol. 6, No.2, Spring 1991, p.158

3. Ibid.; p.158

4. Ibid.; p.158

5. Ibid.; p.155

6. John M. Elliott, *The Official Monogram US Navy & Marine Corps Aircraft Color Guide Vol. 1 1911–1939.* Monogram Publications, Massachusetts, 1987, p.76

7. Although this paint (Naval Gray) was called an enamel, it did not have the high gloss we associate with enamel today, but rather a matte to semi-gloss finish.

Colors and Markings

Standard Colors

Methuen notations are given for known fabric samples, Pantone® equivalents (a printing color standard available from computer suppliers) are given in brackets. Actual colors varies from aircraft to aircraft based on weathering and paint batches; therefore, precise color values can only be approximated.

Color	Methuen	Pantone®
US Roundel Red	10C–D8	(1805C)
	9D8	(1807C)
	9E8	(1807C)
US Roundel Blue	22–23D4	(3015C)
	23C–D8	(307C)
	23B/C6	(2915C)
Italian Roundel Red	9B8	(1795U)
	10C8	(186U)
	10D8	(187U)
Italian Roundel Green	26E8	(357C, 555C)
DH-4 Khaki	5F4; 5F6	(1405C, 147C)
DH-4 Gray (Yellowish)	3/4C4; 3B/C3	(616/617/5865C)
Clear-Doped Linen	2–3A3–4, 3B7	(127C, 110C)
	4A3, 4C3	(155C, 4525C)
Naval Gray	4B2–3	(4545C)

French Five-Color Camouflage

Chestnut Brown	6F 5–7	(497U)
Dark Green	29F3–6	(5605U)
Light Green	30D–E4–6	(3995U)
Beige	4–5D–E6	(132U, 139U)
Ecru	4–5C–D3–4	(451U, 4515U)

Color Plates

1. The Army's 1904 "Gillespie" Medal of Honor design.

2. The Navy's 1913 Medal of Honor design.

3. Macchi M-5, serial number M7229, as flown by Landsman for Quartermaster Charles Hammann, NAS Porto Corsini, Italy, August 21, 1918. Hammann landed M7229 next to Ensign Ludlow's damaged flying boat, allowing Ludlow to escape certain capture. Unfortunately, the cockpit would not accommodate both men, so Ludlow crawled in behind Hammann – under the engine – and laid down flat on his stomach, hanging onto the struts of the engine mount. Both men returned safely to Porto Corsini, over 60 miles away.

The M-5 was a small single-seat fighter which first appeared in early 1918. It was fully aerobatic with a top speed of 118 mph. The wings were constructed of wood and wire bracing, then fabric covered and coated with a protective varnish. The plywood hull was also covered with a protective varnish. But the bottom surface of the boat hull and sides up to the water line were typically painted with a marine paint – usually white. The interplane bracing consisted of tape-bound wooden V struts, with wire-braced tubular metal struts supporting the 160 hp Isotta-Fraschini V4B engine and its pusher propeller.

Although Ludlow's Macchi (M13015) was elaborately painted by late August 1918, most of Porto Corsini's flying boats remained in basic colors until the end of hostilities. Adding the fact that no photographs of M7229 have been found, it is best to illustrate Macchi M7229 in basic markings because the machine would have been delivered to the U.S. Navy this way.

Therefore, the wood hull would have been varnished with light colored marine paint (probably white) applied up to the water line. The serial number "M7229" would have appeared in black characters on each side of the hull, near the bow.

The wings would have been plain linen with the top wing's undersurface segmented into an Italian tri-color pattern, with the red taking up a portion of the port wing, the green a portion of the starboard wing, and the remainder being plain linen. It is possible the center segment (plain linen) was painted white.

Roundels of green, white, and red (outer ring to center) would have been applied to the top wing's uppersurface; the sides of the hull near the cockpit, and the hull underside, near the bow (front). The rudder would have been finished in green, white, and red stripes (forward to aft).

4. Macchi M-5 flying boat, serial M13015, flown by Ensign George H. Ludlow, Porto Corsini NAS, Italy, August 21, 1918. The hull's top section (bow to cockpit) and sides (bow to a point just beyond each roundel) have been painted black. The hull's bottom surface was painted with white marine paint to a point reaching the lower bow roundel. This same white paint was applied on both sides of the varnished wooden hull. Narrow bands of varnished wood remained visible, creating alternating bands of white paint and varnished wood. The outer and inner sides of the wing-tip floats were also painted white.

The wings were plain linen covered. The top wing's undersurface is segmented into the Italian tri-color pattern, with the red taking up a portion of the port wing, the green a portion of the starboard wing, and the remainder being plain linen. It is possible the center (plain linen) segment was painted white.

Roundels of green, white, and red (outer ring to center) were applied to the top wing's uppersurface, the sides of the hull near the cockpit, and the hull underside near the bow (front). The rudder is finished in green, white, and red stripes (forward to aft).

The name of Ludlow's personal aircraft – *MUTT 2nd* – appeared in red on at least the starboard side of the hull.

5. Burgess N-9, serial A-2481, as flown by Francis Edward Ormsbee, Naval Air Station (NAS) Pensacola, Florida, September 25, 1918.

On April 6, 1918, "Naval Gray" was specified the standard color for U.S. Navy aircraft to simplify painting. As with any paint, its primary purpose was to prolong the life of the aircraft by preventing the warping or corrosion typically encountered with salt water. However, the Navy quickly determined that "Naval Gray" greatly reduced the aircraft's visibility, thereby minimizing the chance of detecting the machine in flight. Clearly this was a critical issue with the Navy, because directives entitled "Instructions For Finishing Naval Aircraft," were issued in May and July of 1918 that indicated aircraft would be painted Naval Gray Enamel overall, including among other things, wings, fuselage or hull, floats, struts, exterior spars, and all metal fittings.[7]

The Navy Department issued General Order 299, on May 19, 1917, in its first attempt to address the use of a national insignia on naval aircraft. This Order described the insignia as "...a five-pointed white star inside a blue circumscribed field, with the center of the star red." However, many objected to this design, including Colonel William "Billy" Mitchell, commander of the AEF's Air Service in France, who suggested instead that US aircraft have roundels, similar to the national insignia in use on other Allied aircraft, but with a distinct sequence of colors, thereby standardizing the basic design on all allied aircraft.

Apparently agreeing, the Navy Department issued General Order 364, on February 8, 1918, officially replacing the "star" insignia with a roundel of red, white, and blue (outside to center). This Order further stated no U.S. Naval aircraft was to display the "star" insignia. Although this requirement was adhered to in the war zone in Europe, it was not strictly enforced in the United States. Therefore, many older machines never had their "star" insignia convered to the roundel.

Officially, the United States Air Service roundel was to have a diameter of five feet, when painted on aircraft having a wing chord of five feet or greater. In cases where the wing's chord was less than five feet the diameter of the roundel equaled the chord's length. The diameter of the blue circle was to be two-thirds of the diameter of the red circle while the white circle was to be one third the diameter of the red circle.

Navy department General Order No.299, issued on May 18, 1917, specified that the building number be placed on each side of the rudder, at the top of the white panel, in three-inch high black characters. These numerals were typically applied in larger format to each side of the fuselage and, in some cases, to the center section of the top wing.

The General Order of May 18, was modified by another on August 4, which directed various commands to place prefix the letter "A" before the building number (to denote an aircraft). It appears this requirement was carried out quickly on aircraft rudders. However, those numbers appearing on the fuselage and top wing rarely conformed.

Although no photographs of Burgess N-9, serial A-2481 have been found, basic markings can be determined by the General Orders stated above. A-2481 would have been Naval Gray overall, and roundels would have most likely been applied to the top wing's uppersurface and lower wing's undersurface. This assumption is made on photographs showing roundel-finished, Burgess-built N-9 aircraft that carried lower aircraft building numbers than A-2481.

The aircraft's rudder would have been painted in bands of red, white, and blue (forward to aft), with the white panel displaying the aircraft's build number "2481," with prefix "A," in black.

The aircraft's build number "2481" would have also appeared on each side of the fuselage in large black characters. Although some N-9s also carried their build number on the top wing's center section, its application to A-2481 would be speculative.

On May 21, 1918, the Chief of Naval Operations granted permission for NAS Pensacola, Florida to place squadron markings on seaplanes used for training. However, these marks had to be removed if the aircraft was transferred. It is not known if A-2481 carried such markings.

6. Nieuport 28 serial N6169 "1" as flown by First Lt. Edward V. Rickenbacker of the 94th Aero Squadron. He obtained victories 3–6 in this machine during May 1918. Rickenbacker's third victory – a German Albatros – was obtained on May 17. Moments later, while he pulled N6169 "1" up from a dive, a majority of the top wing's fabric pealed off. Although the plane maintain some level of control, Rickenbacker was lucky to bring it down for a forced landing. That night, squadron ground crews repaired damage sustained in the landing and replaced its wings. The next day N6169 "1" was ready for operations. Rickenbacker would score three more victories in it.

This aircraft was finished in a French five-color camouflage pattern typical to Nieuport 28s. The pattern used chestnut brown, dark green, light green, beige, and black. The undersurfaces were ecru. Early Nieuport 28s issued to the 94th Aero Squadron – March to early May 1918 – came from the factory with French markings. These aircraft had roundels of equal size and position of red/white/blue (outside to center) on the upper and lower surfaces of the top wing, as well as the lower surface of the bottom wing. The French roundels were typically repainted to the American roundel of red/blue/white (outside to center). Likewise, rudders were repainted in white/blue/red stripes (forward to aft). Although this aircraft had its serial numbers overpainted, it should be noted that on some aircraft they were reapplied to the rudder's new white panel.

The 94th Aero Squadron's Hat-In-The-Ring is one of the most famous unit insignias in U.S. aviation history. According to Rickenbacker: "Major Huffer suggested Uncle Sam's Stovepipe Hat with the Stars-and-Stripes for a hatband. And our post Surgeon, Lieutenant Walters of Pittsburgh, Pa., raised a cheer by his inspiration of the "Hat-in-the-Ring." It was immediately adopted and the next day designs and drawings were made by Lieutenant John Wentworth of Chicago, which soon culminated in the adoption of the bold challenge painted on the sides of our fighting planes…"[1]

As the designer, it can be assumed that Wentworth's drawing of the insignia became the standard. Typically, the hat band had six white stripes and five red stripes. The outer stripe was white on both sides of the hat. It had seven white stars in varying degrees of completeness. The inside of the hat band was dark, presumably brown or red, while the deep-shadowed inside of the hat was presumably blue. The brim of the hat was white and the large elliptical ring was red.

Although "two templates were probably used to make the hat-in-the-ring insignia, one for the hat, which left an empty space for the ring, and one for the ring,"[2] a close inspection of the unit's aircraft reveals differences in the angle and placement of the insignia on each machine. Further differences in the insignia are noted in the number of stripes, the coloring sequence of the stripes, and the completeness of the stars. Yet, despite the differences between insignia on an aircraft-to-aircraft basis, it seems the insignia that appeared on the right and left side of each plane matched.

Nieuport N6169 "1" was used on a regular basis by Major John Huffer – the first CO of the 94th Aero Squadron. Following regulations set forth for American squadron, this machine was decorated with a numeral "1" to denote the CO's regular aircraft.

As a general rule, most 94th aircraft carried a white numeral with a narrow dark outline or shadow (in red or black) on the top wing's upper left surface, and a dark numeral (in red or black) on the bottom wing's lower right surface. The white numeral "1" that appeared on each side of the fuselage was further decorated with shadowing in red and blue.

To further differentiate First Pursuit Group aircraft on a squadron level, while at Toul, France, each unit began to apply a stripe of distinctive color and design to their machine's top right and bottom left wing. The 94th adopted a diagonal two-color stripe which was equally divided between red and white.

Additional decorations on N6169 "1" included a painted cowling which appeared in United States Air Service (USAS) roundel colors of white/blue/red (fore to aft).

7. Nieuport 28 serial N6159 "12" as flown by First Lieutenant Edward V. Rickenbacker of the 94th Aero Squadron. Although the Nieuport 28 could mount two machine guns, a temporary shortage of available weapons forced the 94th Aero Squadron to initially equip each machine with only one gun. Despite this, Rickenbacker managed to obtain two victories in N6159 "12", his first on April 29 and his second on May 7, 1918.

This aircraft was finished in a French five-color camouflage pattern typical to Nieuport 28s. The pattern used chestnut brown, dark green, light green, beige, and black. The undersurfaces were ecru.

Appearing to following the general rule, N6159 "12" carried a white numeral "12" with a narrow dark outline (shown in red) on the top wing's upper left surface and a dark numeral "12" (shown in black) on the bottom wing's lower right surface.

Shortly after Rickenbacker's second victory, and after he was put in command of the squadron's 1st Flight, the cowling of N6159 "12" was painted white and a Liberty Loan poster was applied to of the top and bottom right wings.

The poster was designed by Howard Chandler Christy for the Third Liberty Loan Drive. It depicted an attractive young lady unfurling the American flag and carried the slogan "Fight or Buy Bonds…" These posters were likely attached to Rickenbacker's aircraft as part of a publicity campaign. Yet they have become some of the first examples of "nose art" carried by American aircraft during World War One. "The poster came in two sizes: one sheet, 30" x 39½" and half sheet, 20" x 30."[3] But Rickenbacker's machine – N6159 "12" – "carried the one sheet size on both the top wing and the lower wing."[4]

An additional decoration on N6159 "12" was a medallion mounted near the cockpit's starboard side. Although a Saint Christopher's Medal has been suggested, the history behind the medallion remains unknown.

8. Nieuport 28 serial N6283 "16" as flown by First Lieutenant Edward V. Rickenbacker of the 94th Aero Squadron. On May 29, 1918, N6283 "16" was assigned to Rickenbacker for ten days. On June 7 this machine was transferred to Lt. Walter W. Smith. This aircraft was finished in a French five-color camouflage pattern typical to Nieuport 28s. The pattern used chestnut brown, dark green, light green, beige, and black. The undersurfaces were ecru.

All Nieuport 28s starting with serial number N6201 were supplied from the French factory with the correct American markings. As a result N6283 "16" had the Nieuport logo and serial number on the rudder and the standardized red/white/blue (forward to aft) stripes. In addition, roundels did not appear on the undersurface of the top wing. Surviving fabric samples from Nieuports within this serial range "indicates that the white color was noticeably tinted blue and the blue color was exceptionally light (Methuen 22/D/4…"[5]

Appearing to follow the general rule, N6283 "16" carried a white numeral "16" with a narrow dark outline (shown in red) on the top wing's upper left surface and a dark numeral "16" (shown in black) on the bottom wing's lower right surface. Additional wing decorations included the squadron's diagonal two-color stripe of red and white which appeared on the upper right and lower left wing.

As stated, the non-standardization of insignia creates a fingerprint for positively identifying a particular plane. Unfortunately, no clear photograph of N6283 "16" shows this aircraft's insignia. Nevertheless, we can reasonably assume the two templates were used to create the basic design on the aircraft's fuselage.

9 & 10. SPAD 13 C1 (Kellner-built) serial number S4523 "1" as flown by First Lieutenant Edward V. Rickenbacker of the 94th Aero Squadron. Rickenbacker obtained victories 7–26 in this machine. It was finished in a five-color

camouflage pattern typical of Kellner-built SPADs. The camouflage pattern used chestnut brown, light green, dark green, beige, and black. Undersurfaces and wheel covers were ecru. Due to battle damage and normal wear the markings on this aircraft went through several changes over time; the rudder's serial numbers changed in style and position at least twice. The six white dots are battle damage patches on the fin, wing, and fuselage; each carried a small black cross. A white numeral "1" and the squadron insignia appeared on both sides of the fuselage, the hat pointing forward on each side. The outer wheel covers were blue and contained a white star with red dot. The top wing's upper right surface carried a white numeral "1" with a narrow dark outline (shown in red), while the bottom wing's lower left surface carried a numeral "1" (white outlined in red). The squadron's diagonal stripe, of red and white, was carried on the top wing's upper right surface and bottom wing's lower right surface. Additional markings included a red cowling; tri-colored bands – red, white, and blue – which wrapped completely around the forward landing gear struts; and USAS roundels which appeared on the wings in four positions.

Plate 9 shows SPAD "1" as it appeared in the summer of 1918. The rudder carried stripes of blue/white/red (fore to aft). No battle damage patches appeared on the aircraft, and the exhaust pipes were long.

Plate 10 shows SPAD "1" as it appeared in the fall of 1918. The rudder carried stripes of red/white/blue (fore to aft). The aircraft carried six battle damage patches as shown in the color plates, and the exhaust pipes were short. The overall camouflage pattern remained the same as did the other squadron markings.

11. SPAD 13 C1 (Blériot-built) serial number S15202 "26" as flown by First Lieutenant Frank Luke of the 27th Aero Squadron, circa September 1918. There is only photograph that shows Luke with an aircraft that shows some recognizable markings. Although it was not unusual for personnel to be photographed next to some plane other then their own, it more often than not signified ownership. This suggests that Luke flew SPAD No.26 with the 27th Aero Squadron.

The serial number of this aircraft is unknown. However, a list of SPAD XIIIs assigned to the 27th Aero Squadron was compiled by historian Alan Toelle from information in the National Archives. Unfortunately, there is no corresponding list of squadron numbers or assignment of pilot. Yet Toelle indicates that only 19 were Blériot-built, and of these, all but six could be eliminated on the basis that they were either on hand after Luke's death, or not assigned until after that time. Toelle indicates a further group of 29 Blériot-built SPAD XIIIs were allocated to the First Pursuit Group, but their squadron assignment is unknown. He believes eight of these machines were most likely assigned to the 27th Aero on or about July 31, 1918. Thus, the serial number has a one-in-fourteen chance of being correct. Choices include 15155 and 15222, both being assigned to the 27th on September 1. Other equally valid choices would be 15236, 15265, or 15310. This illustration shows 15202, which was assigned on 31 July.

This aircraft was finished in a five-color camouflage pattern typical of Blériot-built SPADs. The camouflage pattern used chestnut brown, light green, dark green, beige, and black. Undersurfaces and wheel covers were ecru. This machine would have been delivered in correct American markings – roundels on top and bottom wings and rudder stripes of red, white and blue (forward to aft), with SPAD logo and serial number.

The squadron's insignia – an eagle imposed over a red circle – appeared on each side of the fuselage, the eagle facing forward. The eagle was gray, with white crest and black wing undersides. The claws and beak appeared in yellow. Typically, the overall configuration appeared consistent from plane to plane, with the principal variation being which parts of the eagle were outlined or highlighted in white.

SPAD S15202 "26", like all aircraft from the 27th Aero, was further distinguished by a diagonal stripe that appeared on the top wing's upper left surface and the bottom wing's lower right surface. The stripe was divided into four bands of alternating black and white checks.

This aircraft carried a black numeral "26" with a narrow white outline on each side of the fuselage, and again, on the top wing's upper right surface and bottom wing's lower left surface.

SPAD S15202 "26" also had an X painted on its engine support structure. This type of marking was found only on Blériot-built SPADs and probably was a quick way to identify the engine manufacturer.

The aircraft also seems to have had a homemade panel installed on the left side of the engine. This was not painted in the normal brown color and could have been plain aluminum, tinplate, or dull gray color.

American pursuit squadrons were organized into three flights and the planes were normally numbered in ascending order according to flight. The flights were further identified by having the radiator shell painted red, white, or blue. SPAD S15202 "26" appears to have a blue radiator shell which would designate the 3rd flight and would be consistent with the number 26.

12. SPAD 13 C1 (Adolphe Bernard-built) serial number S7984, as flown by First Lieutenant Frank Luke of the 27th Aero Squadron on September 28, 1918. Luke shot down three balloons – victories 16, 17, and 18 – only moments before he was forced down and subsequently killed in action. Based on research by historians Royall Fry and Alan Toelle, records of the First Depot at Colombey-les-Belles indicate that S7984 was assigned to the 27th Aero Squadron on September 28, 1918. Since it would be virtually impossible to apply any squadron markings – insignia or numerals – one can reasonably conclude that S7984 bore no squadron markings or numeral when it was lost. It would have been finished in a five-color camouflage pattern typical of Bernard-built SPADs. The camouflage pattern used chestnut brown, light green, dark green, beige, and black. Undersurfaces and wheel covers would have been ecru. USAS roundels would have appeared on the wings in four positions, and the rudder would have had the standardized red/white/blue (fore to

aft) stripes, with SPAD logo and serial number on the white panel.

13. DeHavilland 4 (Dayton-Wright built), numeral "2" (serial number 32169), as flown by Lieutenants Erwin Bleckley and Harold Goettler of the 50th Aero Squadron on October 6, 1918.

This aircraft was finished in a two-color pattern typical of Dayton-Wright DH-4s: the upper wing and fuselage surfaces were finished in khaki, and the so-called yellowish-gray color was applied to the remainder of the fuselage, lower wing surfaces, wheel covers, and wing and landing gear struts.

The squadron's insignia – the Dutch Cleansing Girl, carrying a broom – appeared on each side of the fuselage, the girl facing forward. The girl's dress being blue, with white apron, hood and stick. The shoes appeared in brown. Typically, the overall configuration appeared consistent from plane to plane, with the principal variation being which parts of the insignia were outlined in black.

Photographs of DH-4 "2" show the aircraft both with and without unit insignia, but the dates of the photos are unknown. A memorandum issued by the office of Chief of Air Service on May 6, 1918, indicated that no observation or bombing squadron was allowed to place any distinctive insignia upon its planes until "after one month of service at the front; or immediately upon receiving citation in orders from higher authority for distinguished services." The 50th Aero flew its first mission over enemy lines on September 12, 1918 (during the St. Mihiel Offensive), so it is unlikely the unit's "Dutch Girl" insignia appeared on the Bleckley/Goettler DH-4 (numeral "2") aircraft during the flight of October 6, 1918. Nevertheless, it is shown with it applied.

Additional squadron markings consisted of a double chevron applied to the top of the upper left wing. The apex of the chevron pointed inboard, its two bands making an angle of about 60 degrees to the wing's leading and trailing edge. The width of each chevron appears to equal the width of an outer ring on the wing roundels. The space between the two chevrons appears to equal half the width of one chevron. The inner band was a very light color (probably a light blue). The outer band was a dark color (probably red).

Squadron identification numbers were carried in a dark color (probably red) outlined in white on each side of the fuselage. The number was repeated on the top of the upper right wing and on the bottom of the lower right wing as required by the memorandum issued by the Office of Chief of Air Service, on May 6, 1918.

14. DeHavilland 4 (Dayton-Wright built), numeral "6" (serial number 32517), as flown by Lieutenants Erwin Bleckley and Harold Goettler of the 50th Aero Squadron on October 6, 1918. This was the second and final aircraft used by the Bleckley/Goettler team on October 6, 1918; both aviators were shot down and subsequently killed while attempting to locate the "Lost Battalion."

The overall details of this aircraft are the same as aircraft "2" (color plate 13).

15. DeHavilland 4 (Dayton-Wright built), Day Wing identification number (D-1), as flown by Gunnery Sergeant Robert Robinson and Lieutenant Ralph Talbot, Marine Corps Squadron "C,' Northern Bombing Group, October 14, 1918.

This aircraft was finished in a two-color pattern typical of Dayton-Wright DH-4s: the upper wing and fuselage surfaces were finished in khaki; and the yellowish-gray color was applied to the remainder of the fuselage, lower wing surfaces, wheel covers, and wing and landing gear struts.

Shortly after arriving in France, Major Cunningham, USMC, suggested a distinctive insignia be created and applied to aircraft of the NBG's Day Wing. "The design chosen was made by Quartermaster Sergeant John J. Engelhardt, wing Camoufleur and Sergeant James E. Nicholson, Wing Administration Chief."[6] It consisted of an American roundel of red, blue and white (white center and red outer), imposed over an anchor painted in shades of brown and white. Standing on top of the roundel is an American eagle with wings unfurled.

This is the oldest approved unit insignia in U.S. Naval aviation history. It was typically applied to each side of the fuselage, just aft of the rear cockpit, with the anchor facing forward. Although some insignia appeared on aircraft with the anchor facing aft (dragging), this did not appear the normal practice.

The Marine insignia is shown on aircraft D-1, but it is not known if it was painted on before October 6. Typically this insignia was applied to an aircraft's fuselage after the machine completed its first mission, but D-1 had been assigned to an RAF squadron until early October 1918, leaving little time for such details.

Additional markings consisted of a letter-number identification system that appeared on each side of the vertical stabilizer in light colors (presumably white). Following this system, all DH-4s received by the Day Wing squadrons were designated by the letter "D" and the numbers 1 to 17. The DH-9As were designated by the letter "E" and numbers 1 to 21. The numeral indicated the sequence in which the machines were received by the Wing for operational use in France.

DH-4 Aircraft for the Northern Bombing Group were obtained from the Army as well as the Navy. Several arrived in Miami, Florida, as early as June 1918. When the Marines received DH-4s from the Army, the aircraft's Army serial number was, in many cases, removed from the rudder's white panel and replaced by the Navy serial number. However, while in France, this practice was not strictly enforced. Some aircraft retained their Army numbers and several displayed both on their rudders.

Aircraft D-1 is shown with its Army numbers (32413), but it is not known if its Navy serial number (A-3295) was applied before the flight of October 6, 1918.

1. The Army's 1904 "Gillespie" Medal of Honor Design.

2. The Navy's 1913 Medal of Honor Design.

Jean Galli

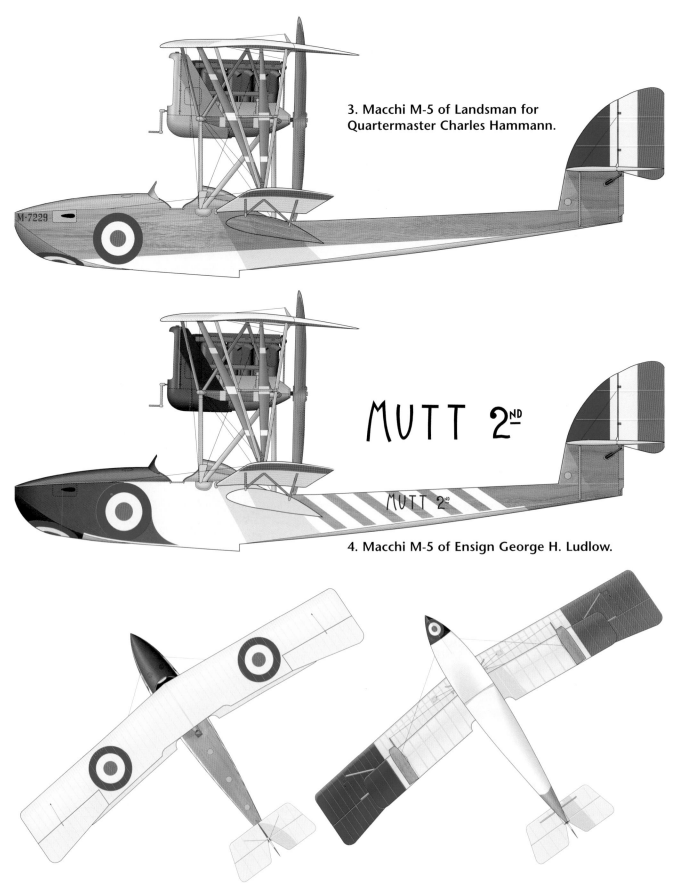

3. Macchi M-5 of Landsman for Quartermaster Charles Hammann.

M-7229

4. Macchi M-5 of Ensign George H. Ludlow.

MUTT 2ND

MUTT 2ND

© Juanita Franzi

4A & 4B. Top and bottom views of Macchi M-5 of Ensign George H. Ludlow.

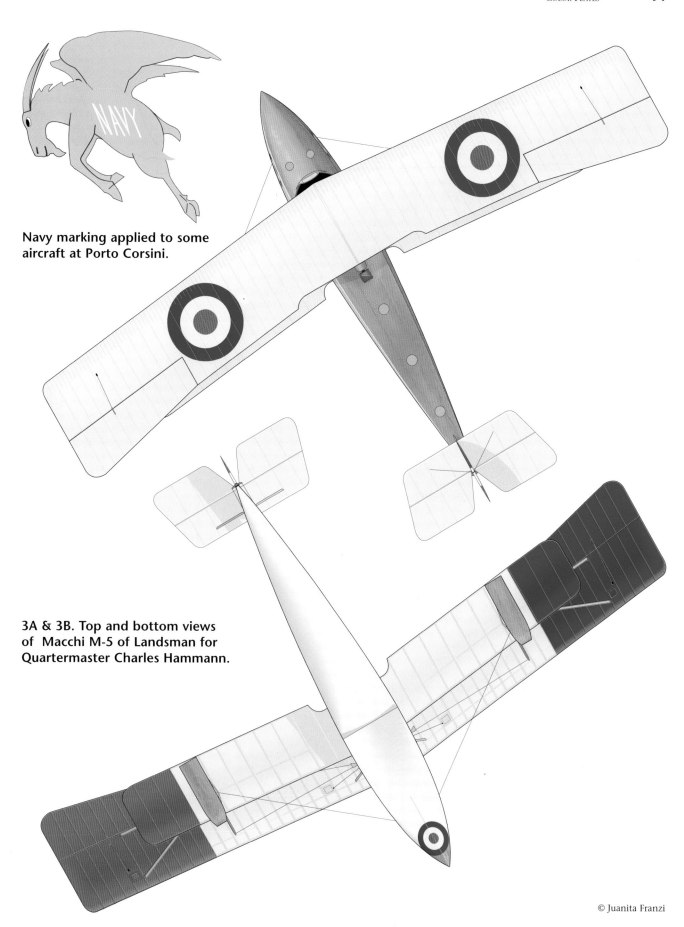

Navy marking applied to some aircraft at Porto Corsini.

3A & 3B. Top and bottom views of Macchi M-5 of Landsman for Quartermaster Charles Hammann.

© Juanita Franzi

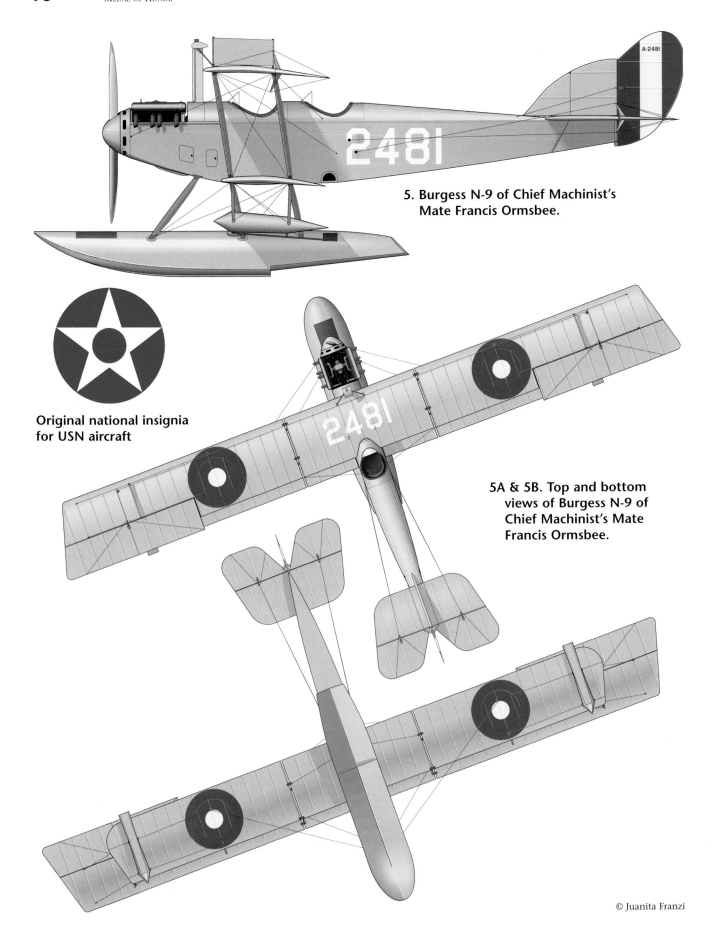

5. Burgess N-9 of Chief Machinist's Mate Francis Ormsbee.

Original national insignia for USN aircraft

5A & 5B. Top and bottom views of Burgess N-9 of Chief Machinist's Mate Francis Ormsbee.

© Juanita Franzi

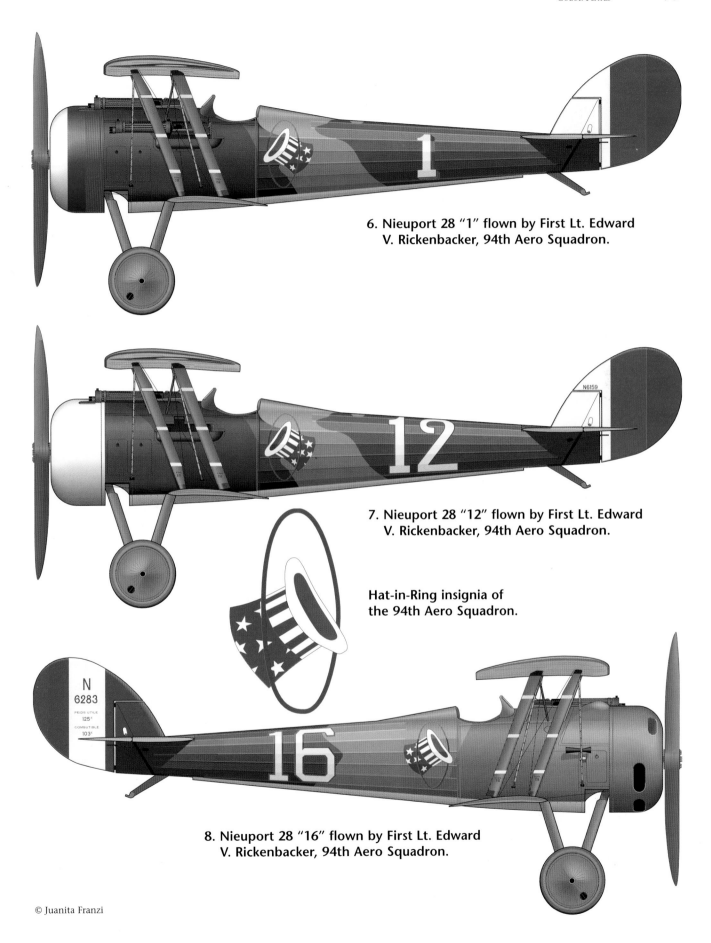

6. Nieuport 28 "1" flown by First Lt. Edward
V. Rickenbacker, 94th Aero Squadron.

7. Nieuport 28 "12" flown by First Lt. Edward
V. Rickenbacker, 94th Aero Squadron.

**Hat-in-Ring insignia of
the 94th Aero Squadron.**

8. Nieuport 28 "16" flown by First Lt. Edward
V. Rickenbacker, 94th Aero Squadron.

© Juanita Franzi

7A & 7B. Top and bottom
views of Nieuport 28 "12"
flown by First Lt. Edward
V. Rickenbacker, 94th
Aero Squadron.

FIGHT
OR
BUY BONDS
THIRD
LIBERTY LOAN

Liberty Bond poster
applied to Nieuport 28
"12" in two locations.

© Juanita Franzi

8A & 8B. Top and bottom
views of Nieuport 28 "16"
flown by First Lt. Edward
V. Rickenbacker, 94th
Aero Squadron.

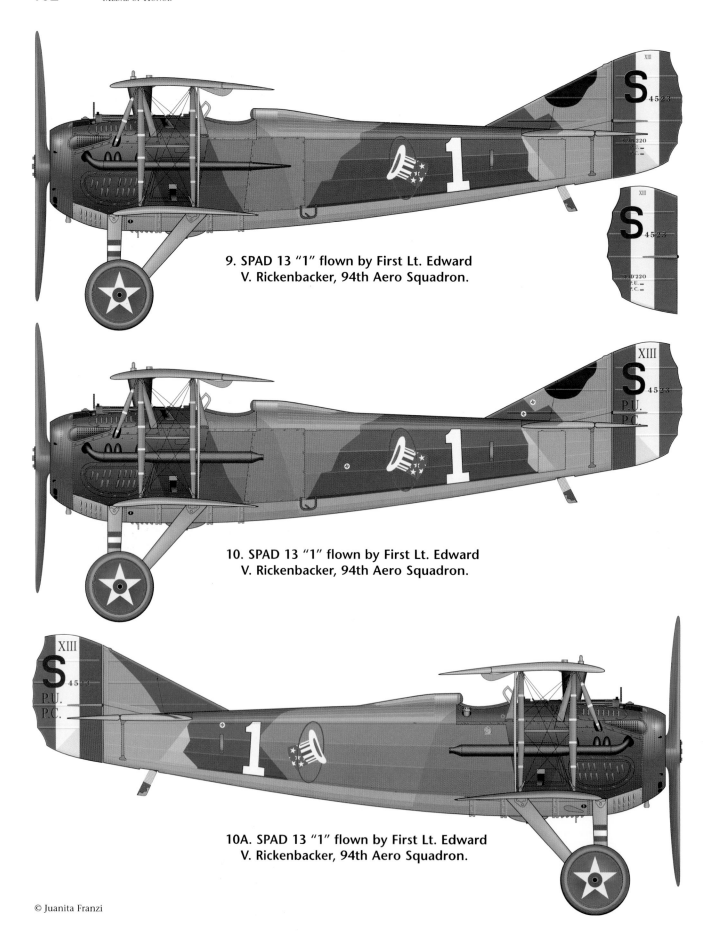

9. SPAD 13 "1" flown by First Lt. Edward
V. Rickenbacker, 94th Aero Squadron.

10. SPAD 13 "1" flown by First Lt. Edward
V. Rickenbacker, 94th Aero Squadron.

10A. SPAD 13 "1" flown by First Lt. Edward
V. Rickenbacker, 94th Aero Squadron.

© Juanita Franzi

10B & 10C. Top and bottom views of SPAD 13 "1" flown by First Lt. Edward V. Rickenbacker, 94th Aero Squadron.

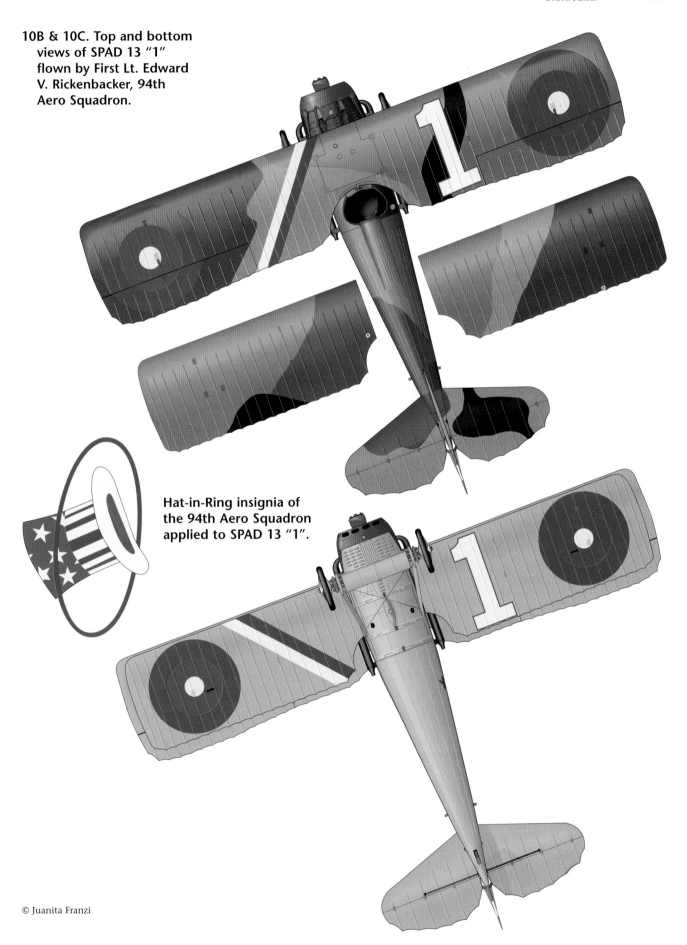

Hat-in-Ring insignia of the 94th Aero Squadron applied to SPAD 13 "1".

© Juanita Franzi

11. SPAD 13 "26" flown by First Lt.
Frank Luke, 27th Aero Squadron.

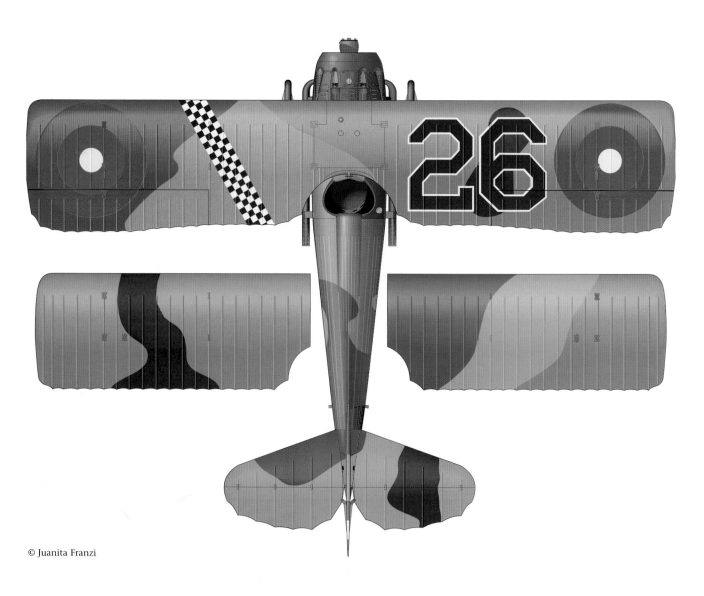

© Juanita Franzi

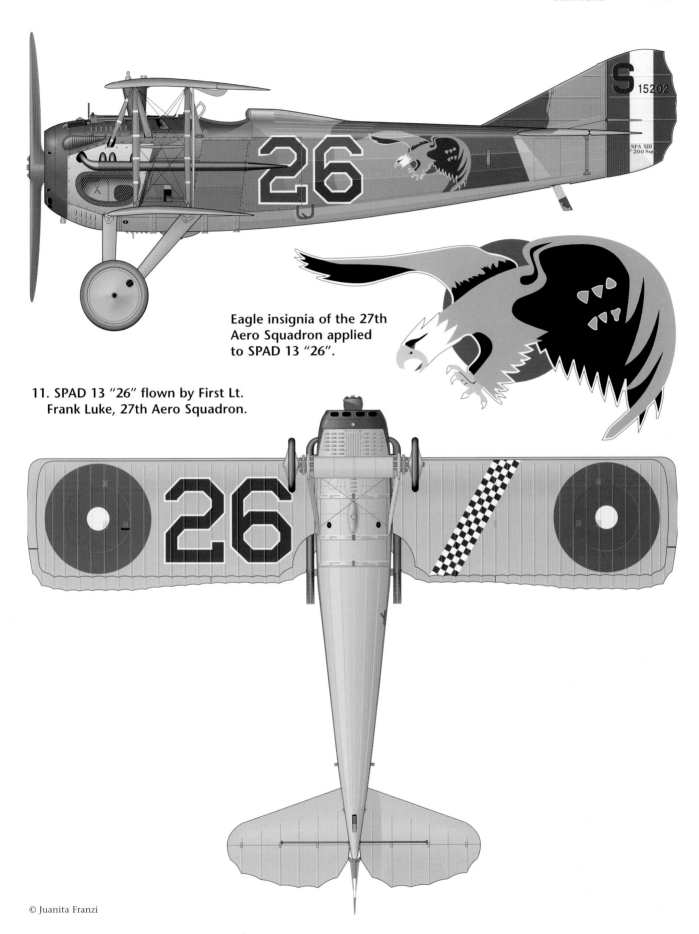

Eagle insignia of the 27th
Aero Squadron applied
to SPAD 13 "26".

**11. SPAD 13 "26" flown by First Lt.
Frank Luke, 27th Aero Squadron.**

© Juanita Franzi

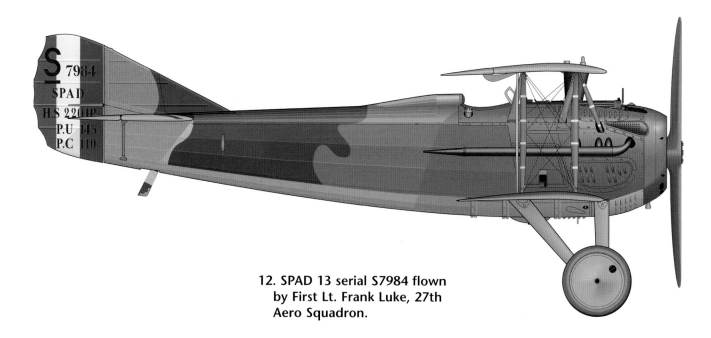

12. SPAD 13 serial S7984 flown
by First Lt. Frank Luke, 27th
Aero Squadron.

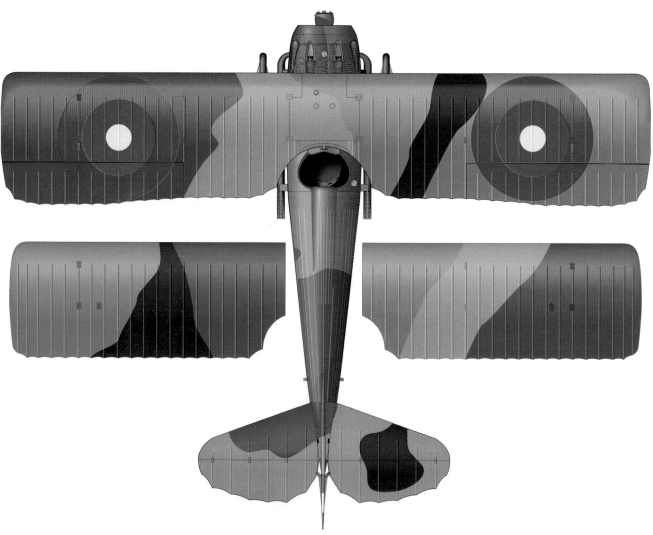

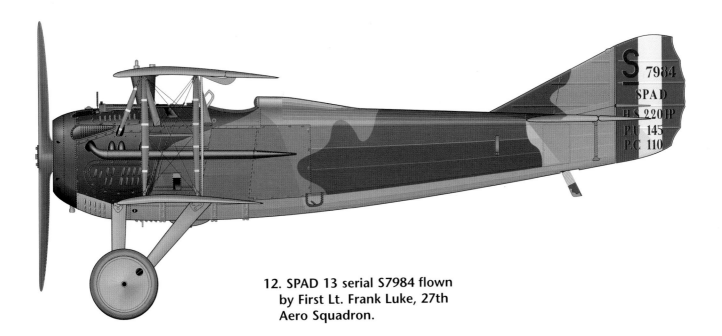

S 7984
SPAD
H.S. 220 HP
P.U. 145
P.C. 110

12. SPAD 13 serial S7984 flown
by First Lt. Frank Luke, 27th
Aero Squadron.

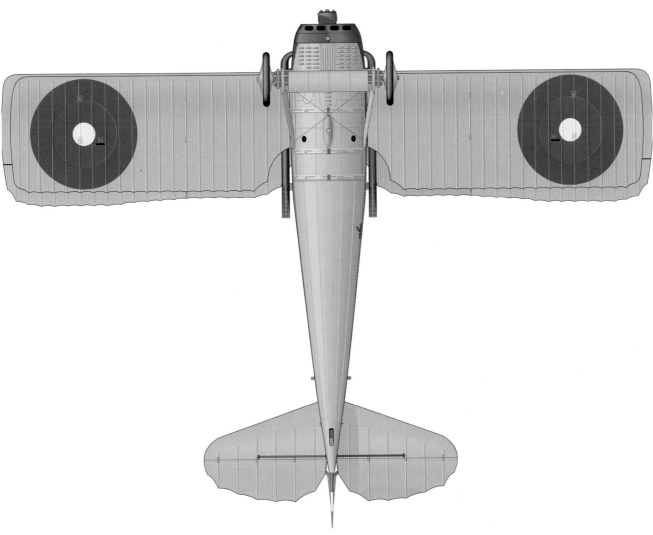

© Juanita Franzi

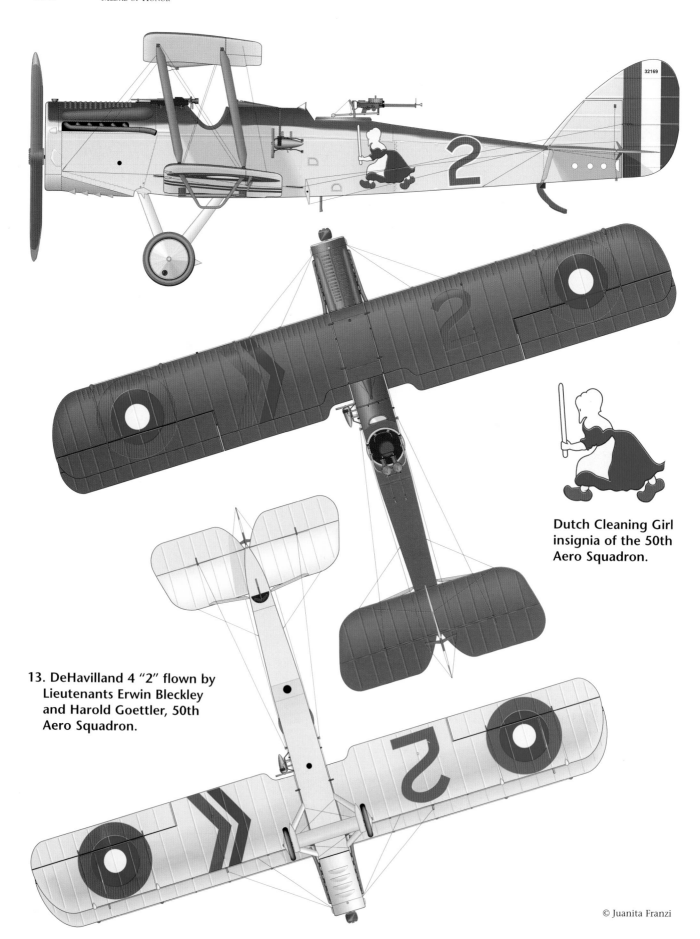

Dutch Cleaning Girl
insignia of the 50th
Aero Squadron.

13. DeHavilland 4 "2" flown by
Lieutenants Erwin Bleckley
and Harold Goettler, 50th
Aero Squadron.

© Juanita Franzi

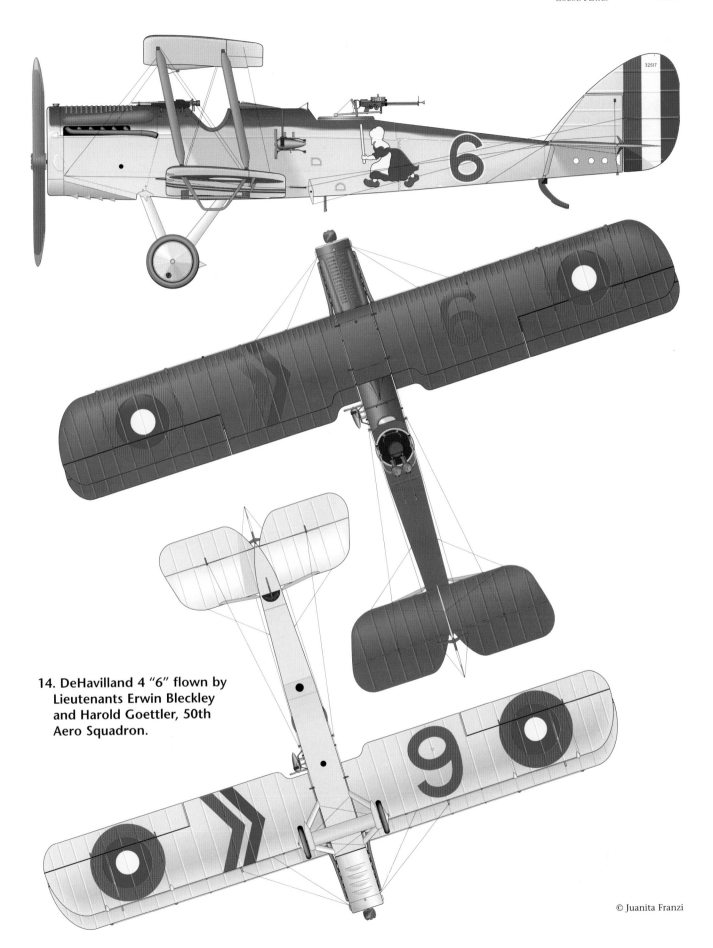

14. DeHavilland 4 "6" flown by
Lieutenants Erwin Bleckley
and Harold Goettler, 50th
Aero Squadron.

© Juanita Franzi

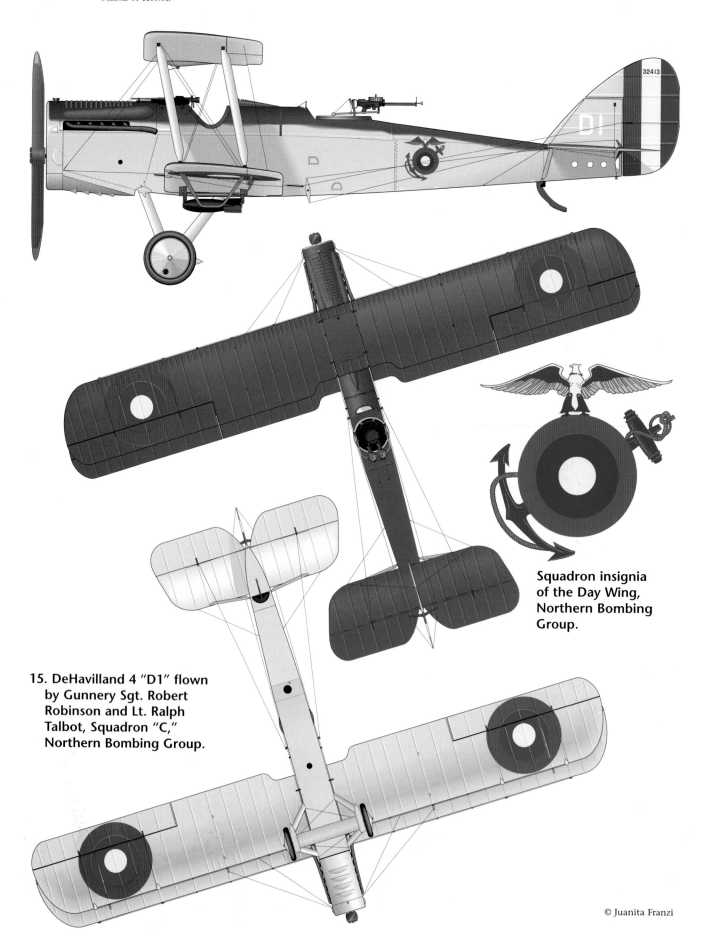

Squadron insignia
of the Day Wing,
Northern Bombing
Group.

15. DeHavilland 4 "D1" flown
by Gunnery Sgt. Robert
Robinson and Lt. Ralph
Talbot, Squadron "C,"
Northern Bombing Group.

© Juanita Franzi